FIXING YOUR DAMAGED
AND INCORRECT GENES

Other World Scientific Titles by the Author

---

*New-Opathies: An Emerging Molecular Reclassification of
Human Disease*
ISBN: 978-981-4355-68-1

A Biography of Paul Berg: The Recombinant DNA Controversy Revisited
ISBN: 978-981-4569-03-3
ISBN: 978-981-4569-04-0 (pbk)

*Emperor of Enzymes: A Biography of Arthur Kornberg, Biochemist and
Nobel Laureate*
ISBN: 978-981-4699-80-8
ISBN: 978-981-4699-81-5 (pbk)

Learning About Your Genes: A Primer for Non-Biologists
ISBN: 978-981-3272-61-3
ISBN: 978-981-120-829-4 (pbk)

FIXING YOUR DAMAGED AND INCORRECT GENES

ERROL C. FRIEDBERG

University of Texas Southwestern Medical Center, USA

World Scientific

NEW JERSEY · LONDON · SINGAPORE · BEIJING · SHANGHAI · HONG KONG · TAIPEI · CHENNAI · TOKYO

Published by

World Scientific Publishing Co. Pte. Ltd.

5 Toh Tuck Link, Singapore 596224

USA office: 27 Warren Street, Suite 401-402, Hackensack, NJ 07601

UK office: 57 Shelton Street, Covent Garden, London WC2H 9HE

Library of Congress Cataloging-in-Publication Data

Names: Friedberg, Errol C., author.

Title: Fixing your damaged and incorrect genes / Errol C. Friedberg,
    University of Texas Southwestern Medical Center, USA.

Description: New Jersey : World Scientific, [2019] | Includes index.

Identifiers: LCCN 2019022655 | ISBN 9789811200960 (hardcover) |
    ISBN 9789811202063 (pbk)

Subjects: LCSH: DNA repair.

Classification: LCC QH467 .F756 2019 | DDC 572.8/6459--dc23

LC record available at https://lccn.loc.gov/2019022655

British Library Cataloguing-in-Publication Data

A catalogue record for this book is available from the British Library.

For any available supplementary material, please visit
https://www.worldscientific.com/worldscibooks/10.1142/11298#t=suppl

Printed in Singapore

Acknowledgements

As was the case when writing the predecessor to this book entitled **Learning About Your Genes**, this book called **Fixing Your Damaged And Incorrect Genes** was written in the hope that individuals interested in contemporary biology might learn something about a biological discipline called **DNA Repair**, a topic that incorporates the many ways that our cells attempt to cope with dangers to our health — and sometimes even to survival consequent to the damage that the DNA in our cells is exposed to. I spent much of my academic career dedicated to research in this fascinating area of molecular biology.

Another sentiment that carried over from **Learning About Your Genes** that bears repeating is that contrary to what one might expect as an author, writing for non-biologists is in many ways considerably more challenging than writing for my peers.

I thank the individuals who read **DNA Repair**, many of whom went to the trouble of providing criticism and advice, and especially wish to thank Sook Cheng Lim and her colleagues at World Scientific Publishing for their patience and efforts in magically transforming a manuscript into a handsome book.

Much of the content of this book was taken from an earlier and more advanced volume entitled, **Correcting the**

Blueprint of Life — An Historical Account of the Discovery of DNA Repair Mechanisms published by Cold Spring Harbor Laboratory Press. I thank John Inglis, director of the press for logistical assistance and for permission to quote extensively from that book.

About the Author

 Errol Clive Friedberg, now retired, was a biologist and historian of science in the Department of Pathology at Stanford University and subsequently served as chairman of the Department of Pathology at the University of Texas Southwestern Medical Center at Dallas, Texas. He studied medicine at the University of the Witwatersrand in Johannesburg, South Africa and received postdoctoral training in biochemistry and pathology at Case Western Reserve University in Cleveland, Ohio before joining the faculty at Stanford University and subsequently the University of Texas Southwestern Medical Center.

Friedberg has written several editions of the textbook *DNA Repair and Mutagenesis*, published by ASM Press, some with multiple co-authors. And he has published several volumes on aspects of the history of molecular biology, including:

Correcting the Blueprint of Life — An Historical Account of the Discovery of DNA Repair Mechanisms.
The Writing Life of James D. Watson.

From Rags to Riches — The Phenomenal Rise of the University of Texas Southwestern Medical Center at Dallas.

Sydney Brenner: A Biography.

A Biography of Paul Berg — The Recombinant DNA Controversy Revisited.

Emperor of Enzymes — A Biography of Arthur Kornberg — Biochemist and Nobel Laureate.

Friedberg has also contributed over 400 papers to the scientific literature, mainly on the topic of DNA repair, and is Founding Editor-in-Chief of the scientific journal, *DNA Repair*. He has received several awards, including the Rous Whipple Award from the American Society for Investigative Pathology and the Lila Gruber Award for Cancer Research. Friedberg is an elected fellow of the Royal College of Pathologists and a Fellow of the American Academy of Microbiology. He received an honorary Doctorate of Science in Medicine at his alma mater the University of the Witwatersrand, South Africa and is an Honorary Fellow of the Royal Society of South Africa.

Contents

Chapter 1 Introduction

An earlier book I wrote entitled, *Learning About Your Genes.* *A Primer for Non-Biologists* informs readers that our **genes** are made of a chemical entity called **D**eoxyribo**N**ucleic **A**cid, commonly abbreviated as **DNA**, an acronym known to most, if not all, Individuals. In addition to other functions, genes direct the synthesis (manufacture) of the hundreds of proteins that are required for the myriad functions in our cells. You would certainly benefit from reading *Learning About Your Genes* before reading *this book*. That being said, such preparation is not essential for your comprehension of *Fixing Your Damaged and Incorrect Genes*.

As you will hopefully glean from your reading, fixing damaged and incorrect genes is generally referred to as **DNA repair**. DNA repair is a collection of processes by which a cell identifies and corrects damage to the DNA molecules that reside in all your cells (with the exception of red blood cells, which do not carry DNA), or for that matter any other molecules that reside in the nuclei of other cells.

Both normal metabolic activities and environmental factors such as radiation can cause DNA damage that can result in as many as **a million individual molecular lesions per cell per day**. (A lesion is any damage or abnormal change in the tissue of an organism, usually caused by disease or trauma.

The word is derived from the Latin *laesio* meaning "injury." Lesions can occur in plants as well as animals.

Many of these lesions cause structural as well as chemical damage to the DNA molecule that can alter or eliminate the cell's ability to function normally, principally in the synthesis of the many hundreds of proteins in the cells in your body. DNA damage can also affect the survival of daughter cells that are generated when cells undergo division to generate new cells. DNA repair is consequently constantly active as it responds to damage of the composition and/or structure of DNA. The extent to which cells that have sustained DNA damage can undergo repair is vital to the integrity and function of their genes and thus to the normal functionality of the affected organism.

The efficiency of DNA repair is dependent on many factors, including the cell type, the age of affected cells and the chemical environment outside cells. A cell that has accumulated a large amount of DNA damage, or one that no longer effectively repairs damage to its DNA, can enter several possible states, including **senescence** (an irreversible state of dormancy), cell suicide (known as **apoptosis** or **programmed cell death**), and **unregulated cell division,** that among other complications can result in the appearance of tumors.

DNA damage due either to environmental factors or abnormal biochemical processes that transpire inside cells, has been estimated to occur at a rate of 10,000 to 1,000,000 molecular lesions (damaged molecules) per cell per day. While this constitutes a tiny fraction of the human genome's approximately 6 billion bases (crucial chemical entities in DNA explained in the next paragraph and in a later chapter on the composition and structure of DNA), unrepaired lesions in critical genes such as those that functionally subdue the formation of tumors (called **tumor suppressor genes**), can impede a cell's ability to carry out its normal function and appreciably increase the likelihood of tumor formation and of various genetic diseases.

The vast majority of DNA damage directly affects the structure of DNA and hence its normal functions. **As discussed in more detail in a later chapter**, DNA contains chemical entities called **bases.** The order and type of the bases (of which there are 4 different types) in DNA constitute a code called the **genetic code**, a code that determines the type and order of amino acids that make up the many hundreds of proteins in your cells, tissues and organs. A simplified explanation of the genetic code and how it works is given in the book *Learning About Your Genes — A Primer for Non-Biologists* mentioned earlier.

Many types of DNA damage result in chemical modification of the DNA bases. These modifications can in turn disrupt the structure (and function) of the many genes contained in DNA by introducing abnormal chemical bonds or bulky adducts (chemical additions). The multiple types of damage that cells can be afflicted with are summarized in the following paragraphs.

DNA damage can be subdivided into two main types: **endogenous damage** such as attack of chemical components of DNA by molecules produced during normal metabolic processes or errors generated when DNA is copied, or **exogenous damage** such as that caused by external agents such as ultraviolet radiation from the sun, other types of radiation (including X-rays), certain plant toxins, various human-made chemicals, or viruses.

Tumor suppressor genes (also called antioncogenes) **encode genes** that make proteins called **tumor suppressor**

protein that **help** control cell growth. Mutations (changes in DNA) in **tumor suppressor genes** may lead to tumors. Tumor suppressor genes are normal genes that slow down cell division, repair DNA mistakes, or inform cells when to die (a process known as **apoptosis** or programmed cell death). Multiple tumor suppressor genes have been identified as being responsible for hereditary cancer.

Consider endogenous damage generated in DNA when that molecule is copied. When cells divide, thereby generating new cells during the growth of organs and tissues as a tiny embryo grows to become a fetus, then a newborn infant, and eventually a fully grown person, DNA molecules are copied by a particular enzyme, a process referred to as **DNA replication**. New cells also arise at any age during the healing of wounds, and to replace the loss of cells that occur normally, a good example being the loss and replacement of skin cells and the cells lining our intestines.

A great many biochemical reactions in our bodies are supported by a special class of proteins called **enzymes** that act as catalysts of the biochemical reactions. The replication of DNA to generate new DNA molecules is supported by an enzyme called **DNA polymerase**, which uses each of the **two strands** of the existing DNA molecule as templates. **RNA polymerase** copies a **single** stand of DNA and generates a single-standard RNA molecule. Most enzymes, including DNA polymerase, are highly accurate in the reactions they catalyze. During DNA replication the newly synthesized DNA is chemically identical to the DNA molecule copied by the enzyme. However, no enzyme that we know of is one hundred percent error-free and occasionally mistakes are made during the process of DNA replication such that the chemical structure of the DNA is altered. (Your understanding of DNA replication and how mistakes during process transpire will be significantly enhanced in a later chapter.)

The rare mistakes that arise during DNA replication are called **mutations**. Mutations can affect the normal functions of genes. However, not all mutations have negative effects. Notably, mutations are the raw material of the genetic variation essential for evolution. Additionally, some mutations may have no demonstrable effect on the health or functions of the bearer of such DNA, in which case they are referred to as **silent mutations**. You, me and all other life forms on planet earth likely carry numerous silent mutations, some of which may distinguish one person from another, in which case they are referred to as **DNA polymorphisms**.

Whereas a mutation is defined as any change in a DNA sequence that is abnormal (implying that a normal version is prevalent in the population and the mutation changes this to a rare and abnormal variant that may or may not alter the normal function of a gene), a polymorphism is a DNA sequence variation that is common in a population but has no functional significance. Polymorphisms are therefore useful for genetic studies in populations, a topic that is beyond the province of this book.

In addition to replication, DNA normally undergoes other biochemical activities. A notable example, during which two DNA molecules exchange segments of any length, is called **genetic recombination**. DNA recombination involves the exchange of genetic material either between multiple chromosomes or between different regions of the same chromosome. This process is generally mediated by homology; i.e., identical or very similar regions of chromosomes line up in preparation for exchange, and some degree of sequence identity is required.

The importance of genetic recombination for the purposes of this discussion on DNA repair, relates to the fact that mistakes during recombination can also alter the normal structure and function of DNA and are a significant source of DNA damage.

In summary, errors in normal DNA function, either during its copying or during its exchange with other molecules, can result in mutations, a prominent source of naturally occurring DNA damage, in this case "damage" defined by incorrectness in the sequence of bases in a DNA molecule.

Up to this point we have been considering examples of **endogenous DNA damage**, i.e. damage produced without an external cause. But components of DNA molecules may also interact with multiple environmental factors such as ultraviolet (UV) radiation from the sun (and more recently from tanning machines) and particularly with a great many chemicals that have an affinity for DNA. All these events consist of **exogenous** sources of **DNA damage**.

X-rays are another potent source of DNA damage, in this case an exogenous damage. X-rays have been used in clinical medicine and for experimental purposes in physics since their discovery in 1895. But their value to genetics research only became apparent when Hermann Muller, an American geneticist, employed radioactivity to produce mutations in the fruit fly *Drosophila*.

Hermann Muller was a member of a research team at Columbia University who, under Thomas Hunt Morgan, an American evolutionary biologist, geneticist, embryologist, and science author who won the Nobel Prize in Physiology or Medicine in 1933 for discoveries elucidating the role that the chromosome plays in heredity, was interested in the physical and chemical nature of genes. As a consequence, he designed experiments to test the idea that radioactive particles might affect genes and lead to altered genes (mutations).

Beginning in late 1926, while at the University of Texas, Muller subjected male fruit flies (of the species **Drosophila melanogaster**, a frequently used organism for genetic studies) to relatively high doses of radiation and mated them to virgin female fruit flies. In a few weeks time he was able to artificially induce more than 100 mutations in the resulting progeny — about half the number of all mutations discovered in Drosophila over the previous fifteen years! Some mutations were deadly. The effects of other mutations were visible in the offspring but not lethal. As Muller interpreted his results, radioactive particles passing through the chromosomes randomly affected the molecular structure of individual genes, rendering them either inoperative or altering their chemical functions.

Fig. 1.1. Thomas Hunt Morgan.

Fig. 1.2. Hermann Muller.

Muller's insight into genes as individual molecular units was provocative and influential. His research in the artificial genesis of mutations (**mutagenesis**) led to the development of the discipline of molecular biology, a complex and varied topic in biology examined by numerous scientists around the world. In 1946 Hermann Muller received the Nobel Prize in Physiology or Medicine.

The Nobel Prize, in the wake of the atomic bombings of Hiroshima and Nagasaki, focused public attention on a subject Muller had been publicizing for two decades — the dangers of radiation. In 1952, nuclear fallout became a public issue as more and more evidence had been leaking out about radiation sickness and death caused by nuclear testing. Muller and many other scientists pursued an array of political activities to defuse the threat of nuclear war.

When exposed to X-rays that emit **ionizing radiation** (IR), DNA can sustain damage due to breakage of the two strands of DNA, an event called **double strand breaks (DSB)**. Double strand breaks are considered the most dangerous of all DNA damage. If left unrepaired, the resulting chromosome discontinuity often results in cell death. Fortunately, this form of DNA damage is rare since there are no natural sources of ionizing radiation (IR). But its ability to seriously damage DNA is often used to kill cancer cells. Indeed, IR is an effective and commonly employed treatment in the management of some human malignant tumors. Since the ability of IR to control tumors mainly relies on DNA damage, the cell's DNA damage response and repair processes may sometimes hold the key to determining a tumor's response to radiation.

In summary, alterations of the chemical composition of DNA and/or its normal function that occur for any reason, be it a consequence of faulty DNA replication, faulty recombination, or exposure to environmental agents that interact with DNA, are collectively referred to as DNA Damage.

Fortunately for you, me and all the other living creatures (including plants), biochemical mechanisms for correcting damage to DNA emerged during biological evolution. Had the phenomenon of DNA repair not evolved our planet would likely be bereft of life forms! **These mechanisms, collectively referred to as DNA Repair, are the topic of this book.**

As stated in the introduction to *Learning About Your Genes*, "like most intellectual disciplines, biology is a vast field with a vast vocabulary, making it difficult (sometimes impossible) for non-biologists to understand much, if anything, about topics such DNA **R**eplication, DNA **R**ecombination and DNA **R**epair (the three **R's** of DNA metabolism). However, in my considered view there is no *a priori* reason why the vocabulary of biology cannot be translated into common English that any reasonably intelligent reader can comprehended."

Fixing Your Damaged and Incorrect Genes was written to address that goal. Accordingly, I have strived to convey biological concepts and explanations about DNA repair in terms that non-biologists can readily comprehend, emphasizing key words in bold type — essentially, to inform you about **DNA repair** in plain English. Additionally, while in this day and age most people with computers are experienced in "goggling" words and even phrases they are unfamiliar with, based on my own experience explanations from **Google** can sometimes be curt and frustratingly uninformative. In an effort to address this potential problem, the end of the book presents a **glossary** of key words and terms that I hope you will use if needed. While In this regard the glossary includes the names of the many scientists mentioned in the book, information about these individuals is something that Google usually handles in an informative manner.

The study of DNA and genes, including their repair when damaged or compositionally incorrect is part of an intellectual discipline called **genomics** (a topic that deals with the structure and function of genes, including the repair of damaged or altered genes). Genomics is a discipline that is formally distinct from a more familiar subject that you have presumably heard or read about, called **genetics**, a topic focused on the study of **heredity**, essentially how characteristics of living beings are inherited from one generation to the next. However, **genomics** and **genetics** are closely related topics, both words being derived from the Greek *genno*, meaning, "to give birth."

Chapter 2
The Early History of the DNA Repair Field

As mentioned in the previous chapter, the DNA repair field emerged from the discovery that X-rays, ultraviolet (UV) radiation from the sun and a myriad of chemicals can interact with and alter DNA. These discoveries spawned the emergence of a new and distinctive investigative focus in genetics, namely the perturbation of genes by exogenous agents and the scrutiny of the ways in which living cells respond to such perturbations — an intellectual discipline that we now call **DNA Repair.**

This focus emerged briefly in the mid-1930s, but its maturation to the intellectually comprehensive body of knowledge that is now referred to as DNA Repair was sporadic and slow. The field did not attain full experimental clarity until the late 1940s and only achieved formal recognition as a distinctive biological phenomenon in all living organisms in the late 1950s.

In an earlier more advanced book about DNA repair that I wrote for scientists and graduate students on their way to becoming scientists (entitled **Correcting the Blueprint of Life — An Historical Account of the Discovery of DNA Repair Mechanisms**), I commented that in the early 1940s insights about mechanisms by which mutations arise and the way that cells responded to genetic insults **(DNA damage)** primarily suffered from an informational vacuum about the chemical nature of genes. This was before a study published in 1944 revealed that the genetic material is made of DNA. The prevailing dogma until that time was that genes are made of protein. Regardless, the dogma was so entrenched in mainstream biology that when in 1953 the now famous American molecular biologist James Watson (who together with the equally famous British biologist Francis Crick) solved the structure of DNA, many (perhaps most) biologists still believed it.

Many of the issues surrounding the dogma that genes are made of protein, not DNA, are recounted in an engaging book by Horace Judson entitled **The Eighth Day of Creation**. Judson informs the reader that Oswald Avery (the scientist credited with the discovery that genes are made of DNA) was not a geneticist (someone who studies genetics) bent on solving the chemical basis of heredity. He was a physician trained as an immunologist and bacteriologist who worked at the Rockefeller Institute in New York, where he was intent on

Fig. 2.1.   Frederick Griffith.

comprehending a phenomenon established in the late 1920s by another physician Frederick Griffith.

Griffith observed that when a **non-virulent** (harmless to humans) form of a bacterium that caused pneumonia in humans was exposed to a cell extract (cell extracts are the material one obtains when one breaks open cells by some sort of drastic measure, thus allowing the cell contents to be saved and investigated) of the **virulent** (harmful to humans) form of that bacterium it permanently acquired the property of **virulence**. Most intriguingly, this phenomenon, called **transformation** by Griffith, was found to be permanent. Griffith consequently dubbed the entity that prompted the switch from the non-virulent to the virulent state **transforming principle.**

Years later, Oswald Avery a scientist at Columbia University, and his colleagues confirmed Griffith's observations and much to their surprise found that the transforming principle was made of DNA, a conclusion that met with considerable skepticism in the scientific community. Many arguments were raised to support this skepticism, not the least being that the genetic material was widely anticipated to be a complex molecule(s) like proteins, whereas DNA, consisting of just 4 different bases could not explain the then known complexities of genes.

We will never know whether his results generated in Avery immediate disappointment, or the excitement of a major scientific advance. But one can confidently conclude that at the time that Colin Avery performed these experiments he was not deliberately in search of the material of which genes are made. He merely wanted to confirm and perhaps expand on Griffith's observations.

Much has also been made of the fact that the now classic study published by Avery and his colleagues appeared in *The Journal of Experimental Medicine*, the "in-house" journal of

Fig. 2.2.   Oswald Avery.

the Rockefeller Institute where Avery worked, rather than in a fundamental genetics journal where it might have had more immediate exposure to geneticists and basic scientists. Nevertheless, this article secured its proper place in history because it represents the first documented demonstration that genes (the transforming principle) are made of DNA.

Judson also pointed out that "although Avery and his colleagues were willing to risk the assumption that Griffith's 'transforming principle' and DNA were one and the same thing," in their published article they qualified this conclusion with the caveat that *"it is of course possible that the biological activity of the substance described is not an inherent property of the nucleic acid (DNA) but is due to tiny amounts of some other substance adsorbed to it ...."*

Though one can understand the skepticism that was pervasive in the basic science community following Avery's discovery, the importance of his scientific contribution surely merited a Nobel Prize, an accolade with which he was never rewarded.

The historical anecdote concerning the work of Oswald Avery and his colleagues illuminates the profound entrenchment of the protein dogma concerning the chemistry of genes. A less celebrated but illuminating example of this bias can be found in a paper presented by Alexander Hollaender at the 1941 Cold Spring Harbor Symposium. Hollaender, one of the leading figures in the then infant discipline of radiobiology, may justly be acknowledged as the founding father of the DNA repair field.

In a paper written in 1986, entitled *History of Radiation Biology From a Personal Point of View*, Hollaender wrote: *"The early nineteen-thirties were an exciting time for those who were interested in modern physics. -------------. I had many interesting discussions with Warren Weaver. He emphasized time and time again that the biological effects of radiation would become very important, especially since physical and chemical approaches can be used extensively. It was Weaver who first coined the term "molecular biology." He is also widely recognized as an important figure in generating financial support for science in the United States."*

Aside from the notion that genes were made of proteins, other dogmas about the nature of genes in the 1930s and 1940s hindered informed knowledge about DNA damage and repair.

The late Robert H. Haynes, a distinguished Canadian geneticist and biophysicist and polymath (a person of encyclopedic learning), and for many years the Distinguished Research Professor in the Department of Biology at York University, was best known to biologists for his contributions to the study of DNA repair and mutagenesis, and totally distinct from his interest in DNA repair, for his contributions to promoting the concept of **terraforming** (the notion of making a planet earth-like), through his invention of the term **ecopoiesis**, a term that refers to the origin of an ecosystem in the context of space exploration), a topic that Haynes was passionately interested in. Haynes defined ecopoiesis as the "fabrication of a sustainable ecosystem on a currently lifeless, sterile planet."

In 1965, I enjoyed a long and informative discussion with Haynes, during which he suggested that "in the early decades of the 1900s cells were regarded as black boxes and genes were regarded as black beads in the black boxes. It was then widely thought that genes were physically stable; that they were somehow sequestered in cells; that they were immune to attack by environmental agents; and that they replicated *precisely* ----- I wouldn't use the term *accurately*, because the notion of replication of genetic information had not yet come forward."

"Another point," Haynes offered, "is that physiological conditions inside the cell were considered to be warm and friendly, and safe for genes. The existence of a potentially substantial mutagenic burden on living cells was completely unrecognized because chemical agents that altered genes and caused mutations were unknown, and the doses of ionizing radiation from X-rays to result in DNA damage were not present at high levels in normal environments."

As early as 1909, the Dutch biologist, Wilhelm Johannsen (who was the first to use the word ***gene***) described mutations as rare, sudden, discrete events that caused genes to change from one stable state to another stable state. In fact this notion of genes as highly stable entities was so deeply entrenched that when the American geneticist, Hermann Muller set about trying to generate mutations in the fruitfly *Drosophila*, thereby finally providing the world the opportunity of studying mutations in the laboratory, he used X-rays, experiments for which he was awarded the Nobel prize in 1946.

Chapter 3

The Composition and Structure of DNA

This is not the time or place to engage readers in the detailed structure of DNA. But it is essential for your understanding of DNA repair to comprehend something about this topic, a feature that stems from a celebrated scientific achievement announced in 1953 by the American scientist, James D. Watson and his British colleague, the late Francis Crick.

Most DNA consists of two strands of equal length, each of which is composed of linked entities called **nucleotides** of which there are four in Nature called **Adenine (A), Guanine (G), Thymine (T)** and **Cytosine (C)** (Fig. 3.1).

Correct nomenclature demands that A, G, T and C are called **bases**, but are referred to as **nucleotides** when they are individually linked to a sugar called **deoxyribose** and a **phosphate** molecule (Fig. 3.1). However, in the interest of avoiding confusion, other than a structural nuance presented in the next several paragraphs and a single DNA repair mechanism discussed later in the book that requires that A, G, T and C be identified as **bases**, A, G, T and C are sometimes referred to in the book as **nucleotides**. To facilitate your comprehension while reading, consider the terms **bases** and **nucleotides** as strictly comparable unless otherwise informed.

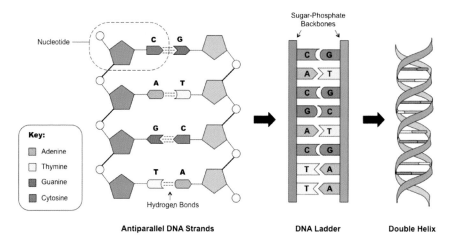

Fig. 3.1. The diagram shows the two DNA helical strands (the double helix as DNA is often referred to) that comprise a DNA molecule. Note that the nucleotide (base) adenine **(A)** is always paired with the base thymine **(T)**, and the nucleotide (base) guanine **(G)** is always paired with the (base) cytosine **(C)**. The **bases** are chemically linked to a sugar-phosphate backbone and exist as **nucleotides**. The two DNA strands exist in an **antiparallel** configuration such that "top" end of one strand is opposite the "bottom" end of the other strand, and are wound around each other as two helical entities, thereby constituting a **double helix**. The term "double helix" is often used when referring to double-stranded DNA.

The absolute length of the two DNA strands varies in different genes. Each strand consists of the four **bases** A, T, G and C, which are attached to a sugar-phosphate backbone, when they are called **nucleotides (hence the distinction between the names bases and nucleotides) (Fig. 3.1)**. The four nucleotides can be in any order in a DNA strand. However, a particular order is unique to any given gene. **No two genes consist of exactly the same order and number of nucleotides in the two DNA strands.**

The organization of the two DNA strands is such that the nucleotide/base **A in one DNA strand** is always paired with **T** in the opposite strand, and **G** is always paired with **C** (**Fig. 3.1**) If for example, one of the two DNA strands of a gene contains a nucleotide/base sequence

ATCTGTTACGC

the opposite DNA strand will have the sequence

TAGACAATGCG

Two bases/nucleotides on opposite DNA strands (such as A-T or G-C) are referred to as **base pairs** and have the identical width. If that were not the case the width of a DNA molecule would vary haphazardly from one nucleotide/base pair to another. Consider how wobbly a ladder would look (and function) if the steps were not of equal width.

Your comprehension of the structure of DNA will be greatly facilitated by carefully reading and understanding the legend to Fig. 3.1.

A simple way of understanding the **antiparallel** configuration of the two DNA strands (mentioned in the legend to Fig. 3.1) is to place two forks next to each other facing in the same direction (**parallel**) and placing one of the forks in the opposite direction (**antiparallel**). You will observe that the width of the two forks remains the same only when in they are in the **antiparallel** position.

DNA in cells is contained in long thread-like structures called **chromosomes** (Fig. 3.2) that reside in a structure in cells

Fig. 3.2. An artist's rendition of two paired **chromosomes**, each comprising two identical **chromatids**. Each chromatid contains a DNA molecule (in red).

called the nucleus. Different organisms have different numbers of chromosomes. In humans, most cells normally contain 23 pairs of chromosomes for a total of 46 and are referred to as *diploid cells*, from the Greek word *diploos*, meaning double. However, sperm in males and ova (eggs) in females, called germ cells, contain 23 *unpaired* chromosomes and are referred to as *haploid cells*, from the Greek word haploos meaning single.

Cells can be opened by various techniques such that the 46 chromosomes are spread out and visualized under a microscope. Individual chromosomes have unique morphology and can be numbered, 1–23. Your cells contain two copies of all your genes. The DNA in one member of a chromosome pair contains a set of gene that you inherited from your mother, and that in the other member of the pair contains a set of genes you inherited from your father.

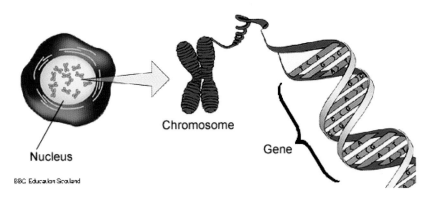

BBC Education Scotland

Fig. 3.3. The cartoon illustrates that **genes** are made of **double-stranded DNA** that resides in **chromosomes, which in turn** reside in the **nuclei** of cells.

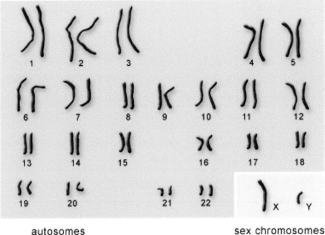

U.S. National Library of Medicine

Fig. 3.4. The cartoon depicts the 23 **autosomal chromosomes** (referred to in the figure as **autosomes**) and the two **sex chromosomes X** and **Y** (bottom right).

The Era of Radiation Biology

Radiobiology (also known as **radiation biology**) is a field of clinical and basic medical sciences that involves the study of the action of ionizing radiation on living things, especially health effects of radiation. Ionizing radiation is generally harmful and potentially lethal to living things but can have health benefits in radiation therapy for the treatment of cancer and thyrotoxicosis, a disease affecting the thyroid gland. Its most common impact is the induction of cancer with a latent period of years or decades after exposure. High doses can cause visually dramatic radiation burns, and/or rapid fatality through acute radiation syndrome. Controlled doses are used for medical imaging and radiotherapy.

As a field of medical sciences, radiobiology originated from Leopold Freund's 1896 demonstration of the therapeutic treatment of a hairy mole using a new type of electromagnetic radiation called X-rays, which was discovered a year previously by the German physicist, Wilhelm Röntgen. After irradiating frogs and insects with X-rays in early 1896 a Russian scientist concluded that these newly discovered rays not only photograph, but also "affect the living function." At the same time Pierre and Marie Curie discovered radioactive polonium and radium later used to treat cancer.

The genetic effects of radiation, including the effects on cancer risk, were recognized much later. In 1927 Hermann Joseph Muller published research showing genetic effects, and as mentioned in Chapter 1 in 1946 was awarded the Nobel Prize for his findings.

More generally, the 1930s saw attempts to develop a general model for radiobiology. Before the biological effects of radiation were known, many physicians and corporations had begun marketing radioactive substances as patent medicine and radioactive quackery. Examples were radium enema treatments, and radium-containing waters to be drunk as tonics.

Marie Curie spoke out against this sort of treatment, warning that the effects of radiation on the human body were not well understood. Curie later died of aplastic anemia caused by radiation poisoning. Eben Byers, a famous American socialite, died of multiple cancers (but not acute radiation syndrome) in 1932 after consuming large quantities of radium over several years; his death drew public attention to the dangers of radiation. By the 1930s, after a number of cases of bone necrosis and death in enthusiasts, radium-containing medical products had nearly vanished from the market.

In the United States the experience of the so-called Radium Girls (female factory workers who developed oral cancers from painting watch dials with self-luminous paint) (but no cases of acute radiation syndrome), popularized the warnings of occupational health associated with radiation hazards. Robley. D. Evans at MIT developed the first standard for permissible body burden of radium, a key step in the establishment of nuclear medicine as a field of study. With the development of nuclear reactors and nuclear weapons in the 1940s, heightened scientific attention was given to the study of all manner of radiation effects.

The atomic bombings of Hiroshima and Nagasaki resulted in a large number of incidents of radiation poisoning, allowing for greater insight into its symptoms and dangers. Red Cross Hospital Surgeon, Terufumi Sasaki led intensive research into the syndrome in the weeks and months following the Hiroshima bombings. Sasaki and his team were able to monitor the effects of radiation in patients of varying proximities to the blast itself, leading to the establishment of three recorded stages of the syndrome. Within 25–30 days of the explosion, the Red Cross surgeon noticed a sharp drop in white blood cell count and established this drop, along with symptoms of fever, as prognostic standards for Acute Radiation Syndrome. Actress Midori Naka, who was present during the atomic bombing of Hiroshima, was the first incident of radiation poisoning to be extensively studied. Her death on August 24, 1945 was the first death to be officially certified as a result of radiation poisoning (or "Atomic bomb disease").

Chapter 5

The Discovery of DNA Repair: Enzymatic Photoreactivation

Long before the consequences of damage to DNA were comprehended, genetic studies, notably the genetics of organisms that can be readily studied in the laboratory, relied heavily on the existence of **spontaneous mutations**, altered biological functions in the absence of any known damage to DNA. However, it was soon recognized that spontaneous mutations are rare events that do not readily lend them themselves to biological studies. Additionally, their recognition by the observation of diseases or other abnormalities was further hindered by the reality that with the exception of germ cells (**ova** in females and **sperm** in males) that possess **a single copy of each gene**, all other cells contain **two copies of your genes**. With exceptions not considered in this book, damage to and consequent malfunction of one copy of a gene does not typically lead to disease or other abnormalities.

The contributions of Hermann Muller and his students at Columbia University discussed in Chapter 2 lent considerable impetus to the emergence of studies on DNA repair. The ability to manipulate the function of genes by exposure of cells to ionizing radiation (X rays) was rapidly broadened to include exposure to ultraviolet (UV) light (a type of radiation present in sunlight that can be readily mimicked in the laboratory by the use of lamps that emit UV light), and eventually to exposure to chemicals that interact with DNA.

But even following the discovery that genes are made of DNA, other scientific dogmas that delayed the awareness of DNA repair prevailed. Notably, contrary to contemporary knowledge that DNA is intrinsically unstable and is highly reactive with multiple chemicals with which they may come into contact, the prevalent sentiment in the 1930s and 1940s was exactly the opposite. The fundamental importance of genes and their critical biological functions prompted the widespread notion **that they must be highly stable entities.**

In the early decades of the twentieth century it was widely thought that genes are physically stable, are somehow sequestered in cells, and are immune to attack by environmental agents. Additionally, it was suggested that the physiological conditions inside cells were "warm and friendly and *safe* for genes" (a notion that I like to refer to as **geneophilia**!) Furthermore, the existence of chemicals as a potentially substantial mutagenic burden on living cells was completely unrecognized. The now established fact that some chemicals are in fact potent **mutagens** (agents that cause mutations) was not yet established and the doses of exposure to mutagens such as X-rays and UV radiation were very low in normal environments. In fact, the choice of X-rays by Herman Muller as an agent with which to produce mutations in the fruitfly *Drosophila* was primarily based on the notion that genes would be too stable to be amenable to modification with anything less energetic.

When I asked Franklin (Frank) Stahl (one of two prominent scientists who made important contributions to our understanding of DNA replication) why the DNA repair field was so late in coming, he responded:

> "I suspect because of a widespread belief (unspoken I suspect, but amounting to worship) among geneticists that the genes are so precious that they must (*somehow*) be protected from biochemical insult, perhaps by being carefully wrapped. The possibility that genes were dynamically stable, subject to the hurly burly of both insult and clumsy efforts to reverse the insults was unthinkable."

Such was the situation prior to the formal discovery of DNA repair as a biological entity.

In the extended historical presentation about DNA repair, entitled **Correcting The Blueprint of Life — An Historical Account of the Discovery of DNA Repair Mechanisms**, published a little over 20 years ago, I wrote:

> "Were it possible to custom design DNA repair mechanisms, perhaps the simplest and most efficient would be those in which specific alterations in DNA are **directly reversed** by single-step reactions. Ideally, such reactions would be catalyzed by stable monomeric (single) proteins with no requirement for exogenous cofactors."

This was not intended as a prediction; I was then simply commenting on what was already well established in Nature. But I subsequently formally categorized this biological phenomenon as **DNA repair by the direct reversal of damage,** to distinguish it from **DNA repair by the excision of damage** presented in a later chapter. One of these reversal processes, called **enzymatic photoreactivation** (or simply photoreactivation) was the first DNA repair mechanism to be discovered, and satisfyingly to historians of biological evolution is also considered the first DNA repair mechanism that appeared on planet Earth.

Among the four bases (nucleotides) in DNA, designated by the letters **A, T, C** and **G**, thymine **(T)** and cytosine **(C)** belong to a chemical family called **pyrimidines** (Fig. 5.1). The many types of DNA damage that result from the exposure of living cells to UV light includes a form of damage in which two immediately consecutive pyrimidines in a DNA strand undergo chemical modification that results in their joining (Fig. 5.1), an event appropriately referred to as **dimerization.** Consequently, these **photoproducts** (a term that includes all altered components of DNA following exposure to UV light) are called **pyrimidine dimers**, of which there are three types: **T<>T dimers, C<>C dimers** and **T<>C dimers** (Fig. 5.1).

As shown in the figure, correct chemical nomenclature considers pyrimidine dimers as **cyclobutane pyrimidine dimers (CPD),** a nuance that you can ignore. For readers interested in chemistry, cyclobutane is an organic compound with the formula

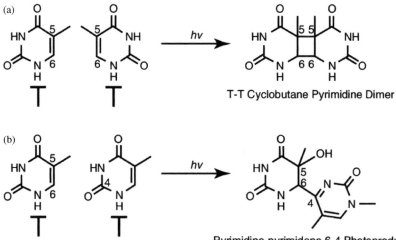

Fig. 5.1. The top half of the figure shows the formation of a T–T pyrimidine dimer from two consecutive thymine molecules in DNA. The bottom part of the figure shows the formation of a 6–4 photoproduct from two adjacent thymine molecules. The symbol **hv** represents visible light.

$(CH_2)_4$. Cyclobutane is a colorless gas commercially available as a liquefied gas. Derivatives of cyclobutane are called cyclobutanes.

Figure 5.1 shows a second photoproduct in DNA exposed to UV radiation, called the **pyrimidine-pyrimidone 6–4 photoproduct (b)**, simplified in casual use to 6–4 photoproducts that can also result from two consecutive pyrimidines in a DNA stand. It is not understood why the exposure of adjacent pyrimidines in UV-irradiated DNA sometimes generates pyrimidine dimers and at other times 6–4 photoproducts.

When pyrimidine dimers and/or 6–4 photoproducts persist in DNA they interfere with its normal functions, notably DNA replication (copying of DNA) as well as another important biological process called **DNA transcription**, during which genes instruct the synthesis (manufacture) of proteins. (Readers interested in how genes in DNA direct the synthesis of proteins are referred to the book, *Learning About Your Genes* mentioned earlier.) In a nutshell, **pyrimidine dimers and 6–4 photoproducts** generated by exposure of DNA to UV light impede fundamental aspects of DNA function.

The discovery of pyrimidine dimers and subsequently of 6–4 photoproducts represent important historical landmarks in the DNA repair field because they informed investigators about the chemical nature of particular types of repairable DNA damage, consequently encouraging further studies on other types of DNA repair. It is no exaggeration to suggest that the multiple mechanisms by which pyrimidine dimers and 6–4 photoproducts in DNA are removed from DNA represent the most extensively investigated of all known cellular responses to DNA damage.

Since, as indicated earlier, UV radiation derives from the sun, the existence of pyrimidine dimers and 6–4 photoproducts is likely as old as DNA itself. Hence, it's not unreasonable to suggest that DNA damage promoted the early evolutionary selection of multiple diverse repair mechanisms to contend with their potentially lethal and mutagenic effects. In fact pyrimidine dimers can be repaired by two different distinct mechanisms, of which **photoreactivation** is specifically dedicated to the repair of pyrimidine dimers.

Two scientists, **Albert Kelner and Renato Dulbecco,** who worked at different research institutions in the United States and who carried out different experiments that led to the same conclusion, **independently discovered photoreactivation.** This coincidence is by no means a fluke. It is a consequence of the unsurprising reality that thought-provoking and attention-grabbing scientific challenges almost always attract the attention of more than one scientist at any given time.

One presumes, perhaps mistakenly, that the general public is largely unaware of the intensely competitive nature of research in most, if not all, fields of endeavor. Sometimes such attention generates collaborative research. Frequently however, research, especially in the life sciences, promotes competition; sometimes with unpleasant consequences for those involved. But one of the primary benefits of competitive research is that it offers the promise of confirmation of the validity of new information.

The goal of historians of science is to garner exclusive first-hand accounts of relevant scientific endeavors. My own aspirations as an historian came close to fulfillment when I was researching this era in the history of DNA repair. As you will presently learn, some of the events surrounding the discovery of DNA repair by enzymatic photoreactivation are as rich in melodrama as a Hollywood movie script.

Adelyn Kelner, wife of the late Albert Kelner, informed me: "My husband was meticulous with his notes and record keeping." This turned out to be a gross understatement. Among other legacies of Albert Kelner's personal and professional life,

Fig. 5.2.   Albert Kelner.

he left over 200 pocket diaries. He also filed away for posterity almost every draft of every major talk he delivered at scientific meetings, and every scientific manuscript he wrote.

In late 1961, a scientist named Claude Stan Rupert (now deceased), a member of the DNA repair community of scientists investigating aspects of DNA repair who went by his middle name Stan, was preparing a comprehensive review on photoreactivation and contacted Kelner for detailed information about his discovery. For reasons that will become clear later in the chapter Kelner was inclined to provide Rupert with a detailed account of the events that led to his discovery; events that he apparently had not shared with anyone except perhaps his wife and closest personal friends.

•••

Albert Kelner came from humble beginnings. Born in 1912 into a poor family, he was stricken with tuberculosis of the bone in his early teens. This affliction required frequent hospitalization and left him with a pronounced and permanent limp. His left shoulder was also affected, a handicap that interfered with his considerable talent as a violinist. His wife Adelyn (with whom I enjoyed multiple poignant and interesting telephone conversations about her late husband) was then of the opinion that had Albert not developed an early interest in biology he would have directed his career to music. "We might have starved as a family," she laughingly confided, "but that's beside the point. He was a very fine musician."

Frequent hospitalization required Kelner to forgo much of his formal high school education. But persistent and diligent informal study gained him a full scholarship at the University of Pennsylvania based on his outstanding entrance examination. Kelner acquired his bachelor, master and doctoral degrees over a period of a mere seven years.

In my several telephone conversations with Adelyn Kelner she spoke of her husband with immense affection. She, not Albert, proposed their marriage! "It was a whirlwind thing," Adelyn related. "I was working in Washington, D.C. at the time and took a brief holiday in upstate New York. I met Albert during the September Labor Day holiday, proposed to him in October and we were married in December."

•••

The late Sir Alexander Fleming was a Scottish physician, microbiologist and pharmacologist. One of his many famous discoveries was the world's first antibiotic substance, penicillin G, from a mould called *Penicillium notatum* in 1928, for which he shared the Nobel Prize in Physiology or Medicine in 1945 with two other microbiologists. Fleming once famously stated:

"One sometimes finds what one is not looking for!
When I woke up just after dawn on September 28, 1928,

I certainly didn't plan to revolutionize all medicine by discovering the world's first antibiotic, or bacteria killer. But I suppose that was exactly what I did."

In the late 1920s Fleming had been investigating the properties of a bacterium called *staphylococcus*, a devastating organism that causes all manner of dangerous infections including severe pneumonia. In early September 1928 Fleming returned to his laboratory, having spent the month of August on holiday with his family. Before leaving his laboratory he had stacked petri dishes (dishes extensively used in biological research laboratories named after the German bacteriologist Julius Petri), on a bench in a corner of his laboratory, on which he had plated staphylococci in order to allow them to grow to colonies visible to the naked eye (Fig. 5.3).

Fleming isolated and propagated the mould and found that it produced a substance that killed a number of disease-causing bacteria. He identified the mould as being from the genus *Penicillium* and after some months of calling it "mould juice", he named the substance it produced **penicillin** — the

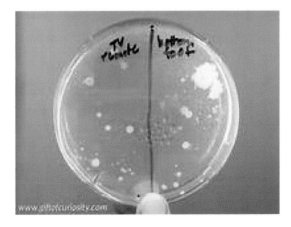

Fig. 5.3. Bacterial colonies growing on agar plates.

Fig. 5.4. Alexander Fleming.

first known antibiotic and an achievement that eventually won him a Nobel Prize!

Alexander Fleming's published discovery of penicillin in 1929 lay relatively dormant in the literature until it was resurrected by a nóted English pathologist, H.W. Flory in the late 1930s.

During WW II, penicillin was mass-produced in the United States and an intense search for other antibiotics was launched. In the years between 1943 and 1946, Kelner was a member of a research team at the University of Pennsylvania involved in this noble quest. His focus was on developing new ways of mass screening bacteria for the production of new antibiotics.

Kelner's efforts gained the attention of Milislav Demerec, a noted Croatian-born microbiologist and director of the Cold Spring Harbor Laboratory, a renowned Mecca for biological research located near New York and at the time of this writing, the domicile of James Watson, co-discoverer of the structure of DNA.

Demerec was deeply involved in the World War II effort to mass-produce penicillin in the US and was keen on exploring the notion that microorganisms might be mutated to forms that excreted new antibiotics.

Supported by funds from a commercial company interested in antibiotic production, Demerec extended an invitation to Kelner to join him at Cold Spring Harbor to screen mutants of a bacterium called *Escherichia. coli* (*E. coli*) for antibiotic production. Kelner once told a colleague: "I thought excretory products might have specific growth-stimulating, growth distorting, or other effects, not only against bacteria but also all sorts of living things. While developing such a method of screening I peddled my ideas around several laboratories hoping they would help me land a job."

Kelner's peddling struck a receptive chord with Demerec, who had assembled a famous collection of bacteriophages (viruses that infect and propagate in bacteria) that had effectively served earlier molecular genetics.

Fig. 5.5. Milislav Demerec.

Horace Judson informed that James Watson remembered Demerec as the man who went around switching off lights, and who refused to make essential repairs like replacing a broken toilet seat!

Upon accepting Demerec's offer Kelner relocated to Cold Spring Harbor Laboratory — a place that like most scientists working there he took to like a duck to water. In order to generate mutants of E. coli that could be examined for the property of excreting antibiotics Kelner exposed the bacterium to ultraviolet (UV) light, an established method of generating mutant bacteria. Accordingly, he exposed E. coli to different doses of UV radiation to establish the amount of exposure that yielded the largest number of mutated bacteria with the least number of killed cells. His notion was to then examine the mutants for the excretion of new antibiotics.

Shortly after beginning his efforts to isolate antibiotic-producing mutants, Kelner was plagued with a recurring quantitative problem. After exposing E. coli bacterium to UV light he repeatedly observed varying numbers of surviving bacterial colonies in different experiments. After multiple experiments executed over a period of several months Kelner concluded that *variations in temperature* in the laboratory that he was working in was the root cause of his inconsistent results.

In later years, Kelner confided to Stan Rupert that "Cold Spring Harbor was very stimulating and I fell in love with microbial genetics immediately, even though the original purpose of my going there to work side by side with a master was not to be achieved. My first task was to irradiate E. coli (a bacterium used in biological experiments for a great many years) with UV light to induce mutants, and from the first experiment in October 1946, I ran into difficulty with the reproducibility of survival rates. … All Demerec could advise was 'more care,' use of as voltage stabilizer, different methods for stirring suspensions, etc. I spent weeks trying to perfect my technique — but to no avail. By October or November of 1946, I had acquired a healthy disrespect for the implications of quantitative exactness of the beautiful UV survival curves in the literature."

By the summer of 1948 Kelner had made little progress in obtaining mutant *E. coli* cells. He was rapidly exhausting Demerec's patience — and more rapidly running out of time before the research grant that supported his work terminated. He decided to make a last ditch effort to establish the nature of what he had concluded were temperature-dependent variations in bacterial survival, by accurately controlling temperature over a narrow range. In order to execute this experiment he moved to a different laboratory located in a different building at the Cold Spring Harbor Laboratory where he could maintain a constant temperature.

Over the course of the next several weeks Kelner's laboratory notebook painfully documented the disaster that had struck his temperature-dependent hypothesis. The results of his new experiments made absolutely no sense to him. He later wrote to a colleague, "I remember feeling very sad, confused and bewildered. The beautiful temperature-recovery curve I thought I had had fallen apart."

However, by early September of 1948 light, in both the metaphorical and literal senses of the word appeared at the end of the tunnel. Illustrative notebook and diary entries at that time stated:

> **Sept. 2, 1948**. Must try the effect of **light-sunlight-diffused light** on recovery …

Sept. 4, 1948. Noted that the 35-degree water bath was **in full light on the lab table**. Noted too that in an old experiment at room temperature, in which recovery was the greatest, the **cells were in transparent bottles on the lab shelf exposed to daylight.**

Kelner's shift in focus to **light exposure** instead of **temperature** as a variable was a fundamental turning point in his experiments. During the course of the next few weeks he systematically examined the effect of light on the recovery phenomenon and obtained unassailable evidence that when UV-irradiated cells were subsequently exposed to visible light they sustained a huge level of recovery compared to cells that were shielded from light exposure. Temperature had nothing to do with the recovery! Focused as he was on the notion of isolating mutants that had acquired the ability to generate and excrete new antibiotics Kelner cannot be faulted for failing to recognize that he in had fact discovered a form of light-dependent repair of DNA damaged by exposure to UV light, **the first experimental evidence of DNA repair**!

Toward the end of his extensive correspondence with Stan Rupert, Kelner wrote:

After the final conclusive experiment — I could delay closing my laboratory no longer. I told Demerec right away about the visible light effect, and although he perked his ears up a trifle and some expression of interest crossed his face, there was little comment. Taking stock of the future, to see what I could do next, and realizing that there was no job prospect, I asked Demerec what was to be done. He said I could stay through the winter and work with Bryson (another scientist at Cold Spring Harbor) on a grant obtained for Bryson on antibiotic resistance. This I refused, for how could I abandon the recovery work now?

I made a bargain with him — let me concentrate on the recovery problem and I would guarantee to leave by May. ---------- Demerec said later that I could have continued with photoreactivation (the name of the DNA repair mechanism that he had unwittingly discovered) — but this is simply not true. The recovery work wasn't mentioned at all. ------------ Others have told me that that I should have accepted Demerec's offer and worked on recovery anyway, deceiving Demerec as to what I was doing. I was too naive for this.

During the stressful period that Kelner was unable to comprehend why, when he exposed cells to UV light, he was confronted with distressing variation in the survival of irradiated cells, he discussed this problem with numerous scientific colleagues, including a highly respected and well known investigator named Salvador (Salva) Luria at the University of Indiana. During this period, Kelner was extremely concerned about finding another position in order to support himself and his family. He was also then aware that a young scientist named Renato Dulbecco worked in Luria's laboratory.

In late October of 1948 Kelner wrote a lengthy letter to Luria in which he explained in detail the events that had transpired during his attempts to generate mutants that excreted new antibiotics. Deeply concerned about retaining his employment at the Cold Spring Harbor Laboratory, he also explained his dire need to find a new job in the fervent hope that Luria could be of assistance in that regard.

About a month later, Kelner received a reply to his letter from Luria.

"Dear Kelner, you will be interested in knowing that Dulbecco has discovered, quite by accident, a phenomenon which may be the counterpart in phage (phage are viruses that infect bacteria that were then and still are

*frequently used in biological studies) of your discovery
on bacteria ----- Ultraviolet-inactivated phage is reacti-
vated by visible light at a terrific rate."*

In essence Luria informed Kelner that a member of his laboratory had observed the same phenomenon that Kelner had. Kelner was stunned by this news and immediately shared Luria's letter with Demerec and others at Cold Spring Harbor. In his correspondence to Stan Rupert 13 years later Kelner penned:

*"Demerec and the staff at Cold Spring Harbor were
far more indignant and skeptical than I, and told me so. ...
Of course photoreactivation would have been discovered
eventually, ----- and maybe even Dulbecco would have. But
he certainly knew about my work before making his observations. You can imagine how I felt at the time, with no job,
or opportunity to work, and anxiety about the future."*

Urged by Demerec, Kelner began writing up his results for publication in a premier science journal: *Proceedings of the National Academy of Sciences*, a journal in which Demerec enjoyed the privilege of publishing without formal review of his scientific articles since he was an elected member of the prestigious US National Academy of Sciences.

Kernal subsequently received a letter from Luria stating:

Dear Kelner,

*"Because of the extreme interest that the photoreactiva-
tion (it would appear that Luria had unilaterally already
named this phenomenon) will have for virologists, we
have thought that Dulbecco should send a note to Nature
(a prominent scientific journal that only a small majority
of scientists manage to publish in) briefly relating the*

facts. I thought that unless you have already published your results you might like to send in a similar note. I am enclosing a copy of Dulbecco's note.

Dulbecco ran into photoreactivation in a most queer manner, by forgetting to put off the fluorescent light on a table on which he had left a pile of UV-irradiated plates. Next day the top plate had 100 times more plaques than the bottom one and the intermediate ones had gradually different numbers. ------ It is a most exciting thing, and I imagine that the bacterial phenomenon you discovered must also be such."

Regrettably I was unable to obtain a firsthand measure of Kelner, an apparently gentle and private man, or to determine what he thought and how he felt during the period between late December 1948 and January 15, 1949, before he replied to Luria's final letters. When he finally did so he composed a masterpiece of professional sobriety and decorum in which he adopted a calm, reasoned and forthright appeal to what was obviously an emotionally charged situation for him.

Several people who recalled that time, notably his wife Adelyn, told me that securing Luria's good graces to help him find a job was not a trivial motive in his demeanor. Yet he was at the same time unwilling to capitulate on the important principle at stake for him — recognition and priority for his years of experimentation.

On January 15,1949 Kelner wrote again to Luria. Two days later, Luria responded. Most of the rest of his lengthy letter is not relevant to this discussion. He ended the missive stating:

"In view of the above, I think it only fair that that you should have the complete credit for the first discovery of

photoreactivation. My suggestions, which I want to sub-mit to you for approval before anything is done (besides stopping the publication of the note in Nature, which I have already done telegraphically) are the following:

a) Dulbecco's note could have the following paragraph inserted after the first one: 'The occurrence of photoreac-tivation of ultraviolet irradiated phage was noticed acci-dentally a few weeks after receiving a personal communi-cation from Dr. A. Kelner that he had discovered recovery of ultraviolet treated cells upon exposure to visible light. My observation indicated the correctness of Dr. Kelner's suggestion that the phenomenon discovered by him may be of general occurrence for a number of biological objects.

b) If you consider this satisfactory, the note on phage could be sent on to the publication if you do not expect to publish you're your discovery soon. --------- If, however, you plan to publish soon Dulbecco agrees to delay pub-lication of his observation until that time."

On January 20, 1949 Kelner responded to this commu-nication stating:

"The solution you suggest is a most fair one and if the insertion and emendation you suggest are included in the letter to Nature I of course give my whole-hearted approval for the immediate publication of Dulbecco's findings.

At Demerec's suggestion I had submitted a manuscript for publication some weeks ago, and perhaps if possible you might want to mention this paper as 'In press, Proc. Nat Acad. Science.' This is not too important a point, and it would not be worth delaying publication of Dulbecco's manuscript to include this reference."

Kelner ended this letter writing:

> *"I'm very glad to have this affair off my mind and look forward to discussing the scientific points of this phenomenon."*

The correspondence between Kelner and Dulbecco was more extensive than that shown in this book. Readers interested in reading the entire Kelner/Dulbecco correspondence are referred to an article published in 1999.

(Friedberg, E. C. *The discovery of enzymatic photoreactivation and the question of priority: The letters of Salvador Luria and Albert Kelner. Biochemie* **81**: 7–13, 1999.)

Kelner's priority for the discovery of photoreactivation was formalized when his article was published in the *Proceedings of the National Academy of Science* several months before the appearance of Dulbecco's letter to *Nature*. The latter publication included all the amendments that Luria had promised.

In the final analysis, there is no doubt that Renato Dulbecco discovered the phenomenon of photoreactivation independent of Albert Kelner's studies. Dulbecco eventually switched his scientific interest to the study of oncoviruses (viruses that promote cancer in experimental animals), studies that earned him the Nobel Prize in Physiology or Medicine in 1975. He spent many years at Caltech, a prestigious university located in Pasadena, California, approximately 11 miles northeast of downtown Los Angeles, and at the Salk Institute in California.

Many years ago, I visited Renato Dulbecco. We met in Francis Crick's opulent presidential office at the Salk Institute in La Jolla, California (named after Jonas Salk who famously

Fig. 5.6. Renato Dulbecco.

Fig. 5.7. Salvador Luria.

developed a vaccine against poliomyelitis) that overlooks the Pacific Ocean as far as the eye can see, where I asked about the tumultuous events surrounding the discovery of photore-activation and its significance in the DNA repair field.

Dulbecco's recollections of his encounter with photoreactivation were understandably vague. Aside from the fact that they had transpired a long time ago, he did not view them as especially significant in it his career development, which soon switched to animal viruses. Hence, the drama recounted in the correspondence between Kelner and Luria was not at all vivid in his memory. When I suggested that, in retrospect, photoreactivation was the lifeblood of Kelner's scientific existence, whereas for Luria it was more of an interesting coincidental thing, he emphatically agreed. "Exactly" he told me "And for me too, because in fact I didn't really pursue it at all."

Dulbecco had absolutely no recollection of any correspondence between Kelner and Luria and expressed no special interest in reading it 50 years later. Indeed, Dulbecco (who died in 2012) frankly admitted that while he certainly did not disdain history, he had no particular interest in the topic.

As for Albert Kelner, there is little doubt that his discovery of what was called photoreactivation resurrected his scientific career about which he was once seriously concerned. Soon after completing his sojourn at the Cold Spring Harbor Laboratory he obtained a special United States Public Health Service Fellowship to work at Harvard University. During that period, Kelner published an article on photoreactivation in the journal *Scientific American*, an accolade accorded to scientists by invitation to draw attention to important biological phenomena. Soon thereafter he was recruited to the faculty at nearby Brandeis University that had just opened its doors and where he spent the remainder of his scientific career.

Following the publication of his initial experiments Kelner had thoroughly searched the literature for documented evidence that photoreactivation might have been previously documented, a task he probably should have undertaken earlier. He discovered that in 1933 two German physicists reported that visible light inhibited the browning action of UV radiation on banana skin, and that in 1941, a scientist at Stanford University reported that white light counteracted the growth-inhibiting action of UV radiation an a particular alga.

• • • • • • •

Kelner's discovery of photoreactivation is considered in many quarters to be serendipitous. The history of science is replete with examples of this quirk of human endeavor. Perhaps the most celebrated is the discovery of penicillin by Alexander Fleming, the very field that in which Kelner began his career.

In fairness to Kelner I do not share the view that his discovery was at all accidental.

The word *serendipity* denotes engagement in a search for a particular objective and totally by chance obtaining an outcome that is very different, and typically not at all unpleasant. I believe that the word owes its origin to an old fairy tale entitled, *The Three Princes Serendip* (the former name of Ceylon).

In this story, the principal characters, three princes in pursuit of a certain princess's hand in marriage, encountered examples of unexpected and unintentional good fortune. The version of this tale that I once read chronicled that one of the princes was searching for a lost arrow and instead encountered a beautiful maiden who he married. Certainly, a very different (and pleasant) outcome from the task of searching for a lost arrow. Judge for yourself whether or not Kelner's discovery belongs in the category of unadulterated chance!

• • • • • • •

Enzymatic photoreactivation represents the first example of the discovery of the repair of DNA damage. The enzyme-catalyzed conversion of pyrimidine dimers in DNA to the correct monomeric forms TT, CC and TC is also the first example of the first known **DNA repair by the direct reversal of DNA damage,** a mode of repair that is completely error-free. In subsequent years, the repair of various types of chemical damage to DNA notably chemicals called **alkylating agents,** were shown to also be reversible. Features of the repair of alkylation damage to DNA are discussed in Chapter 9.

Kelner's original experiments that led to the discovery of photoreactivation were carried out using spores of an organism called *Streptomyces griseus*. He later repeated these experiments using the bacterium *E. coli*, an organism that soon became a favorite model for molecular genetics and molecular biology. In these experiments Kelner noted that when *E. coli* cells were kept in buffer or water in the dark, i.e., in the absence of photoreactivating light, a two- to three-fold increase in survival occurred. Since the cells were scrupulously shielded from photoreactivating light, this recovery, small though it was, reflected a *light-independent* DNA repair mode that was later shown to be excision repair (a phenomenon discussed in Chapter 7).

Kelner was so absorbed with the dramatic light-dependent phenomenon that he missed the opportunity of discovering excision repair. So he failed to pursue the even more provocative observation that when ultraviolet-irradiated *E. coli* cells are put into a favorable medium in the dark for three hours, they lose their ability after to recover from subsequent exposure to photoreactivating light. What transpired of course was that during incubation in the dark the cells repaired most of the DNA damage by excision repair. Hence the loss of photoreactivation!

These results were reported at a meeting in 1949, also attended by Renato Dulbecco and Max Delbrück. Delbrück, who died in 1981, has often been called the founder of molecular biology. In 1969, he shared the Nobel Prize for Physiology or Medicine for work in the area of molecular genetics.

In April 1949, Luria and Dulbecco presented talks at a Meeting held at Oak Ridge, Tennessee. Dulbecco spoke about the discovery of photoreactivation. In a later biography of Max Delbrück entitled *Thinking About Science: Max Delbrück and the Origins of Molecular Biology*, the authors noted that Delbrück made reference to the utility of what he called "the principle of limited sloppiness" with respect to the discovery of photoreactivation. "If you're too sloppy," he wrote, "you never get reproducible results, and you never can draw any conclusions; but if you are just a little sloppy, then when you see something startling you ------nail it down."

Delbrück had previously expressed his amazement that photoreactivation had not been discovered earlier. In late 1948, he wrote to Salvador Luria: "Photoreactivation is a shocker, and it's a miracle that it was not discovered before. It shows that everybody else was working too sloppily to notice it, and you — too precisely to encounter it. It is the old story of the principle of measured sloppiness that leads to discovery."

• • • • • • •

Following the publications by Kelner and Dulbecco it was by no means evident that photoreactivation was a DNA repair process. The phenomenon was independently observed by an Italian scientist at the University of Pavia who quite reasonably suggested that postirradiation exposure to light destroyed "cellular poisons" that were induced by UV radiation. Several historians of photoreactivation have pointed out that the "cellular poisons" theory of UV radiation and its reversal by some sort of light-dependent remained prevalent for some time.

By the early 1950s, Kelner had extensively characterized the phenomenon of photoreactivation. He astutely honed in on the observation that one of the earliest photoreactivable consequences of UV radiation exposure in bacteria was inhibition of DNA synthesis. He consequently concluded, "the first consequence of ultraviolet absorption must be a change in nucleic acid molecules. He hypothesized that:

> "The ultraviolet-induced inhibition of desoxyribonucleic acid (DNA) is correlated with a general change in the nucleus, which results in an inhibition of all or many of the reactions of the cell which are governed by the nucleus — that is, ultraviolet paralyzes nuclear function. Reactivating light removes the paralysis and renders the nucleus functional."

As I wrote in **Correcting the Blueprint of Life**: "History might be generous enough to concede such language is as explicit an assertion **of DNA damage and repair as "damn it" is to swearing!** But the hard facts are that that the notion of **DNA Repair** was not categorically promulgated, and neither Kelner nor anyone else exalted the phenomenon of photoreactivation to this status at that time. In discussions with John Cairns, whose contributions to the DNA repair field are revealed in a later chapter, Cairns stated:

> "I can remember feeling that photoreactivation was not a real proof that DNA lesions can be repaired. My thinking was that if one form of light can make a lesion, it is not surprising that another form of light can undo the lesion."

A fundamental conceptual limitation to understanding the significance of photoreactivation was the lack of definitive evidence that the target for the process was indeed DNA. By the late 1940s, the correlation between Griffith's "transforming principle" and DNA was well accepted, and in retrospect the experimental observation that UV-irradiated transforming factor could be photoreactivated would have provided compelling evidence that the primary cellular target for UV radiation was DNA, not protein. But when photoreactivation was discovered, the only known transformable bacteria were *Diplococcus pneumoniae* (the organism used by Avery and his colleagues) and *Haemophilus influenza*. By some confounding anomaly of evolution, neither of these two organisms are endowed with photoreactivation!

Consequently, even as late as 1950, the notion that cells had evolved specific mechanisms for repairing damaged DNA had not yet penetrated the rapidly evolving world of molecular genetics. However, in 1952 when Salvador Luria, the mentor of Renato Dulbecco was discussing the possible mechanism of a phenomenon called *multiplicity reactivation* (a process by which two or more virus genomes, each containing inactivating DNA damage, can interact with an infected cell to form a viable virus genome), he used the term *DNA repair*. He wrote:

"The occurrence of photoreactivation — without contradicting the hypothesis of localized damage in discrete determinants, suggested the need for caution in interpreting multiplicity reactivation, since physiological mechanisms of **repair** may be involved." This may well have been the first documented use of that word with reference to biological responses to DNA damage.

If one uncovers a novel observation in a living organism one must be able to give it truth by dissecting and comprehending the observation outside the organism! The first definitive demonstration of photoreactivation *in vitro* (outside an organism) and hence the first clear indication of enzyme-catalyzed DNA repair, was published in 1956 based on studies that a young scientist named Claud Rupert carried out in collaboration with his colleague Sol Goodgal in the laboratory of Roger Herriot at Johns Hopkins University.

Rupert had obtained his Ph.D. in physics from Johns Hopkins University in 1951 and remained there as an inexperienced postdoctoral fellow in the newly formed Department of Biophysics. When I commented on the large number of early molecular biologists who were reformed biophysicists, he retorted with a twinkle in his eye: "Yes, just a golden time for that to happen because you didn't have to know anything! You didn't have to know any biology to suddenly stumble into this empty territory!"

"Goodgal made it sound awfully easy," Rupert once told me. "According to him all we had to do was grind up some *E. coli*, centrifuge out the junk and mix the juice with

Fig. 5.8. Claud (Stan) Rupert.

UV-irradiated DNA. Then expose the mixture to light, transform streptomycin-sensitive cells and see whether we get strepto-mycin resistant cells."

Having never done a biochemical experiment, Rupert initially protested that it was too complicated. But with Good-gal's guiding hand, Rupert set up an experiment in which UV-irradiated transforming DNA was incubated with an *E. coli* extract in the presence or absence of visible light. They added ATP and magnesium to the extract because Bob Lehman and Arthur Kornberg at Stanford University had done so in their experiments on DNA replication!

Years later, here's how Rupert described the outcome of the first such experiment. "I put this thing together and Sol looked in every now and then and we put the bacterial Petri dishes in the incubator on a Saturday night. The next morning Sol was in the lab and called me at home and said that one of the plates had 10 times more bacterial colonies than the rest of them and wanted to know which one that was. So I looked at my notebook, that I had with me and told him that was the one in which the DNA had been exposed to light. On Monday morning, I trotted down to the lab and Goodgal asked me where all my stuff was. I pointed to the freezer and showed him where everything was and he told me to give him my lab notebook, with instructions not to touch anything until he was through "I'm going to do it all over again myself," he said. He did and it turned out the same way.

A week after the experiments were completed (June 19, 1956), a symposium called "**The Chemical Basis of Hered-ity**" was convened at Johns Hopkins University. Goodgal was squeezed into the program at the eleventh hour to present his and Stan Rupert's preliminary findings. "I think it took a while

Fig. 5.9. Stan Rupert and Sol Goodgal, Johns Hopkins University (FIRST ROW, 2ND AND 1ST FROM RIGHT). Courtesy of Dr. C.S. Rupert, University of Texas at Dallas. Noncommerical, educational use only.

for the discovery to sink in," he related to me. I was sitting next to Cy Levinthal at the meeting and when I sat down after my talk he gave me all sorts of accolades.

Goodgal's and Rupert's results on photoreactivation in the test tube were published in the proceedings of the conference under the title **"Photoreactivation of Haemophilus influenza Transforming factor for Streptomycin Resistance by an Extract of *E. coli*,"** the first documented evidence that photoreactivation of DNA damage was an enzyme-catalyzed reaction. The field of DNA repair was born and the baby was not at all unattractive to the scientific community.

Fig. 5.10. Aziz Sancar.

On the 21st Anniversary of their 1953 publications on the structure of DNA, Francis Crick revisited the two famous *Nature* papers and conceded that "We totally missed the possible role of enzymes in repair, although due to Claud Rupert's very elegant work on photoreactivation, I later came to realize that DNA is so precious that probably many distinct repair mechanisms would exist. Nowadays one could hardly discuss mutation without considering repair at the same time."

Some years after the discovery of photoreactivation, an enzyme (appropriately called **photoreactivating enzyme**) was discovered by **Aziz Sancar** at the University of North Carolina and was shown to be the agent by which pyrimidine dimers in DNA generated by exposure to UV light are restored to their normal state. The phenomenon of DNA repair then became a recognized biological phenomenon.

6 The Emergence of DNA Repair

Prior to the formal discovery of NER several investigators reported that when UV-irradiated *E. coli* cells are in a growth medium, or even in a buffer solution before being plated on agar plates, their survival improved compared to that of cells plated immediately following exposure to UV light. Since the bacterial survival transpired in a liquid solution this observation was referred to as **liquid holding recovery**.

During that period interest in the biological responses to DNA damage was muted. It was largely the province of a discipline called radiation biology or radiobiology, a field of clinical and basic medical sciences that involves the study of the action of ionizing radiation on living things, especially the health effects of radiation.

As a field of medical sciences radiobiology originated from the demonstration of the therapeutic treatment of a hairy mole using a new type of electromagnetic radiation called X-rays, which had been discovered previously by the German physicist Wilhelm Rontgen. After irradiation of frogs and insects with X-rays in early 1896 it was concluded that these newly discovered rays not only photograph, but also "affect the living function." At the same time Pierre and Marie Curie discovered radioactive polonium and radium later used to treat cancer.

The genetic effects of radiation, including the effects on cancer risk, were recognized much later. In 1927 Hermann Joseph Muller published his research showing genetic effects, and in 1946 was awarded the Nobel Prize.

The 1930s saw attempts to develop a general model for radiobiology. Notable here was Douglas Lea an experimental physicist working primarily in the field of radiobiology whose efforts included an exhaustive review of some 400 supporting publications.

The first death ever to be officially certified as a result of radiation poisoning was termed "Atomic bomb disease."

In the late 1950s the understanding of DNA repair underwent a dramatic change. A Canadian scientist named Ruth Hill working at York University in Toronto isolated a strain of *E. coli* that was abnormally sensitive to killing by exposure to UV light independent of visible light, thereby distinguishing it from photoreactivation and accelerating the mounting conviction that some type of DNA repair other than photoreactivation likely transpires in all organisms. The discovery of this mutant turned out to have a major impact on the ascension of the DNA repair field.

Ruth Hill died suddenly of a cerebral hemorrhage in 1973. In a tribute to her contributions to the DNA field written a year later in the proceedings of a DNA repair conference held in Squaw Valley, Bob Haynes stated:

> "Hill won an important and secure place in the history of biology by virtue of her discovery in 1958 of the first *radiation-sensitive* mutant bacterium strain of *E. coli*. The isolation of this mutant came as a surprise to radiation geneticists, who up to that time had been more concerned with the isolation of *radiation-resistant* strains. Furthermore, the extent to which the mutant affected sensitivity, especially to UV light, was vastly greater than

had ever been observed before for other physical, chemical, or biological modifiers of radiobiological action.

It is not surprising that the very existence of such a mutant should immediately have attracted the interest of radiation biologists, most of whom were accustomed to dealing with three- or four-fold changes in sensitivity but certainly not with sensitization factors of 100 or more. The many theories and experiments stimulated by Ruth Hill's discovery changed the complexion of cellular radiobiology. But even more significantly, comparative studies of the mutant and wild type strains led in 1964 to the discovery of DNA excision repair, one of the few fundamental processes in molecular biology that was not foreseen by the clairvoyant pioneers of that field."

Concerning the distinction between radiobiology and DNA repair, one of the pioneers of the budding field of DNA repair once told me:

"I think it necessary to distinguish between what the radiobiologists thought and what the molecular biology

Fig. 6.1. Robert (Bob) Haynes.

community paid attention to. The radiobiologists generated killing curves and tried to understand what was going on. I believe that until there was some biochemistry attached to these curves none of the molecular biologists engaged in working out the details of the genetic code and of transcription paid the slightest attention. What Dick Setlow (one of the primary pioneers in the DNA repair field) accomplished was to attract mainline biologists to what was going on. In a similar vein Matthew Meselson, then a leading light in the molecular biology of DNA told me that in the early 1960 that "the only radiobiologists able to command some audience outside of the field were Dick Setlow and Alex Hollaender (then a well established radiobiologist.) Many of those people had a very limited knowledge of modern genetics, and that was the key necessary to open the door."

Chapter 7

Excision Repair of DNA

A notable feature of photoreactivation is that during biological evolution it disappeared in most placental mammals (any member of the mammalian group characterized by the presence of a **placenta**). While this may be sad news for those of you who love to bask in the sun, the good news is that unlike photoreactivation, excision repair pathways (**there are multiple types of excision repair**) are repair processes that remove numerous types of damage to DNA.

In contrast to photoreactivation, which is specifically and uniquely directed to pyrimidine dimers but which was evidently lost during the evolution of most placental mammals (mammals that are born with a placenta), excision repair is a much more general and ubiquitous process. The discovery of photoreactivation formally secured the concept of DNA repair in Nature, but the discovery and subsequent elaboration of the full repertoire of excision repair modes established the generality of repair of a large medley of implied lesions. As the term implies, during excision repair chemically altered nucleotides and bases are enzymatically *excised* from the genome and replaced by normal ones.

This theme of base or nucleotide replacement can be effected by an astonishing variety of enzymes that eliminate an equally astonishing array of abnormal bases and nucleotides, not only those damaged by environmental physical or chemical mutagens (compounds that cause mutations in genes), or by spontaneous chemical alterations, but also bases that are mispaired as a result of errors during DNA replication, as well as bases that are inappropriate to the usual chemistry of DNA, such as the presence of uracil instead of thymine.

Nature solved the replacement problem for all excision repair modes by the process of DNA synthesis. Whereas in principle the accurate duplication of genetic information by DNA replication requires only a single-stranded genome, the replacement of nucleotides lost from one strand of double-stranded DNA has an obligatory requirement for the complementary strand to serve as an informational template.

This observation, coupled with the prevalence of UV radiation from the sun as a natural source of DNA damage suggests the intriguing possibility that the theme of base or nucleotide replacement as a strategy for coping with genomic injury, rather than DNA replication may have provided the primary evolutionary selection of double-stranded DNA. Robert (Bob) Haynes, Philip Hanawalt and others in the young DNA repair field enthusiastically subscribed to this evolutionary notion, but the idea was not widely promulgated outside the field.

Regardless, members of the early DNA repair community recalled that although the famous conclusion by James Watson and Francis Crick in their celebrated publication in 1953 announcing the structure of DNA that reads:

"it has not escaped our notice that the specific (base) pairing we have postulated immediately suggests a possible copying mechanism for the genetic material"

was presumably focused on DNA replication, it is equally valid for DNA synthesis during DNA repair. In fact a less celebrated but equally insightful observation in an article published in *Scientific American* by Robert (Bob) Haynes and Philip (Phil) Hanawalt in the mid-1960s stated:

> **"The two strands of DNA are complementary because adenine in one strand is always hydrogen-bonded to thymine in the other, and guanine is similarly paired with cytosine. Thus the sequence of bases that constitute the code letters of the cell's genetic message is supplied in redundant form. Redundancy is a familiar stratagem to designers of error-detecting and error-correcting codes. If a portion of one strand of the DNA helix was damaged, the information in that portion could be retrieved from the complementary strand. That is, the cell could use the undamaged strand of DNA as a template for the reconstruction of a damaged segment in the complementary strand."**

One would have to go back in evolution to the time when DNA synthesis was first selected in some long extinct animal to solve this evolutionary conundrum.

The direct experimental discovery of excision repair reflects the congruent efforts of a number of key players in different laboratories, notably **Richard (Dick) Setlow** and his colleagues Paul Swenson and Bill Carrier at the Oak Ridge National Laboratory located in Oak Ridge, Tennessee; **Paul Howard-Flanders** and his postdoctoral fellow Dick Boyce at Yale University; and **Philip Hanawalt** and his graduate student David Pettijohn at Stanford University. Importantly, however, aside from the significance of their contributions in formalizing the existence of a DNA repair mode that operates in the

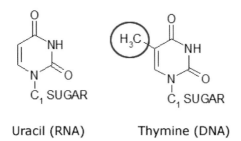

Uracil (RNA)   Thymine (DNA)

Fig. 7.1.   Uracil (found in RNA) and thymine (found in DNA) are chemically very similar except for the presence of a methyl group (CH3) in thymine.

absence of light, the direct experimental elucidation of excision repair demystified the many vague allusions to alternative DNA repair modes generally couched in terms such as "recovery" and "reactivation," and placed this repair paradigm firmly in the arena of modern molecular biology.

The first type of excision repair discovered is one in which damaged bases are excised from DNA as **nucleotides** (bases linked to the sugar deoxyribose and to a phosphate molecule) that are included in small DNA fragments about 20 nucleotides in length. I formally designated this mode of excision repair as **nucleotide excision repair (NER)**. Tomas Lindahl demonstrated that when (as sometimes happens), DNA contains **uracil (U)** instead of **thymine (T)** [the two bases are chemically very similar (Fig. 7.1)], uracil is removed as the **free base** rather than as a nucleotide. This mode of excision repair is now called **base excision repair (BER)**.

Nucleotide Excision Repair (NER)

Prior to the formal discovery of NER, several investigators reported that when UV-irradiated *E. coli* cells are placed in a liquid growth medium or even in a buffer solution before being plated on agar plates, their survival improved compared to that of cells plated immediately following exposure to UV light. Since the bacterial survival transpired in a liquid solution, this observation was (peculiarly) referred to as **liquid holding recovery**.

In the mid-1960s, Paul Howard-Flanders at Yale University and Richard (Dick) Setlow, his wife Jane Setlow and their respective colleagues independently demonstrated that following exposure to UV light, bacteria such as *E. coli* **excised** (cut out) small pieces of **single-stranded DNA** from their genomes (a **genome** refers to the complete collection of genes in a cell) that contained pyrimidine dimers. Howard-Flanders and his colleagues subsequently isolated three mutant strains of *E. coli* that were defective in this process. The mutations in these strains were mapped to different regions of the *E. coli* genome, indicating that proteins generated by at least three discrete genes are involved in the excision process, now called **nucleotide excision repair**. In due course, research in multiple laboratories resulted in the isolation of a collection of distinct proteins that are required for NER.

Fig. 7.2. Richard (Dick) Setlow.

Bob Haynes later recalled that in the fall of 1963 Dick Setlow visited him at the University of Chicago to give a seminar. "I remember him coming into my office — we knew each other because we were both members of the old radiobiology club and the biophysics club — and he triumphantly said to me: "Do you know how *E. coli* repairs UV damage?" I said, "No, how?" He replied, "it cuts out dimers and throws them away," or words to that effect. I realized that was a remarkable experimental observation, but it didn't really hit me as "Wow!" because I had thought of that idea independently and others had thought of it too. But Dick was the first to get the experimental data that proved this without a doubt."

Others had indeed thought about excision repair. In a paper published in 1963, Walter Guild wrote: "An entirely speculative mechanism — is that the presence of damaged DNA acts as a substrate-inducer to elicit an adaptive enzyme formation — namely a specific enzyme. This nuclease removes

Fig. 7.3.   Jane Setlow.

the damaged strands of DNA, along with considerable adjacent regions, preparatory to resynthesis of new DNA from the remaining intact strand."

In a letter he wrote to me in 1995, Bob Haynes provided an interesting summation of the complex history that in his view surrounded the elaboration of excision repair:

"It is possible," he wrote, "the concept of excision repair, at least in rough outline, arose in the laboratories of multiple investigators, including Setlow and myself. What is interesting is that all of us knew one another and attended the same meetings. But because this idea seemed to be so speculative in 1962/63 (before Setlow's experiments), it's possible that none of us was sufficiently excited to mention it to one another. However, I do know that Dick (Setlow) was very excited by the results of his excision experiments, and I became excited when he visited us in Chicago in 1963. It became immediately clear to me that he had critical evidence for this mechanism upon which others merely speculated."

Fig. 7.4. Philip Hanawalt.

Figure 7.5. portrays the key events during NER of DNA damage associated with exposure of cells to ultraviolet (UV) radiation. Unlike enzymatic photoreactivation, during which photoreactivating enzyme uniquely recognizes pyrimidine dimers and 6–4 photoproducts in DNA, NER was soon shown to remove a large number of different types of damage from DNA.

It remains unclear how the NER machinery recognizes multiple types of base damage in DNA. It has been cogently suggested that pyrimidine dimers generated by exposure of cells to UV radiation as well as multiple other types of DNA damage that are repaired during NER are recognized by the DNA repair machinery as localized distortions of the DNA double helix.

In 2013, Philip Hanawalt, a former graduate student of Setlow, and a long-standing friend and scientific colleague during the many years that we were both faculty members at Stanford University, published an informative article entitled *The Awakening of DNA Repair At Yale*. He wrote: *"The experiments that led to the discovery of DNA excision repair were initiated in the Yale Biophysics Department just a few years after James Watson and Francis Crick proposed the double-helical structure for DNA. The complementary strands of that duplex structure immediately suggested to them a mechanism by which DNA might replicate. But the possibility had been overlooked that an intact DNA strand might also serve as a template for the precise **reconstruction of a damaged region in the other strand**."*

Around this time, the study of the effects of UV light on DNA replication led to the discovery by Hanawalt and his student David Peffijohn that DNA synthesis transpires during the excision of pyrimidine dimers. In 1963 Hanawalt and Pettijohn published a preliminary report on DNA replication in which they stated that their results are consistent with the hypothesis that "some type(s) of photochemical damage to DNA may result

NORMAL DNA
ATCAGGCTTATTCGAAACTAGT
TAGTCCGAATAAGCTTTGATCA

EXPOSURE TO UV LIGHT GENERATES
A TT DIMER IN A DNA STRAND
ATCAGGCTTATTCGAAACTAGT
TAGTCCGAATAAGCTTTGATCA

THE DIMER IS EXCISED AS PART OF A
DNA FRAGMENT LEAVING A
GAP IN THE AFFECTED DNA
STRAND
ATCAGGC   TAAACTAGT
TAGTCCGAATAAGCTTTGATCA
+
TTATTCG
THE GAP IS FILLED IN WITH NORMAL
NUCLEOTIDES BY DNA
SYNTHESIS RESTORING
A NORMAL DNA STRAND
ATCAGGCTTATTCGAAACTAGT
TAGTCCGAATAAGCTTTGATCA

Fig. 7.5.   Excision repair of DNA.

in the partial replication of the DNA molecule." In a definitive exposition published a year later, they related this DNA replication to excision repair.

Collectively these observations led to a model of how pyrimidine dimers are removed from DNA by the process now called **n**ucleotide **e**xcision **r**epair (**NER**), the essential features of which are shown in Fig. 7.5.

Transcription-Coupled Nucleotide Excision Repair (TCNER)

Before considering transcription-coupled nucleotide excision repair, it is relevant to understand the differences between

DNA and a closely chemically related molecule called RNA, which is primarily generated during a process called **DNA transcription**.

DNA transcription is a biological event during which a molecule called **RiboNucleicAcid (RNA)** that is chemically similar to **DNA** is copied from a DNA strand, typically during the process of protein synthesis during which RNA is transcribed from the coding strand of DNA.

Structurally **DNA** and **RNA** are nearly identical. However, there are fundamental differences between the two polynucleotides that account for the different functions of DNA and RNA. Notably

DNA contains the sugar **deoxyribose** while **RNA** has the sugar **ribose**.

DNA is stable under alkaline conditions while **RNA** is not.

DNA is usually double-stranded, whereas **RNA** is **single-stranded**.

DNA contains the base **thymine** that pairs with **adenine**.

RNA contains the base **uracil** that pairs with **adenine**.

DNA and RNA also have different functions. **DNA** is responsible for storing and transferring genetic information, while **RNA** directly codes for amino acids and acts as a "messenger" between DNA and a structure called the **ribosome** where amino acids are assembled into proteins. When so engaged, RNA is called **messenger RNA**, a label bestowed by the late Sydney Brenner.

To counteract prolonged blockage of transcription cells remove RNA polymerase-blocking DNA lesions by a DNA repair mode called **transcription-coupled nucleotide excision repair (TCNER)**, a specialized sub-pathway of nucleotide excision repair (NER). During TCNER the RNA polymerase arrested at a lesion constitutes a signal for recruitment of specific TCNER factors and the arrested RNA polymerase is removed or back-tracked to allow access to conventional NER repair enzymes.

TCNER is dedicated to the repair of lesions in DNA that by virtue of their location on actively transcribed DNA **arrest RNA synthesis**, thereby crippling biological events that depend on the presence of RNA, notably protein synthesis. The encounter of an elongating RNA polymerase with DNA lesions can have severe consequences for cells, since this event can generate a signal for the arrest of the normal cell cycle and sometimes even for the programmed cell death of affected cells.

Exposure of mice to UV light or chemicals has demonstrated that TCNER is a critical survival pathway that protects against acute toxic and long-term effects (such as cancer).

Deficient or defective TCNER in humans is sometimes associated with mutations in two genes called **CSA** and **CSB** that can result in a disease called **Cockayne syndrome (CS)**.

CS is a rare disorder characterized by an abnormally small head size (microcephaly), a failure to gain weight and grow at the expected rate (failure to thrive) leading to very short stature, and delayed development. The signs and symptoms of this condition are usually apparent from infancy, and they worsen over time. Most affected individuals have an increased sensitivity to sunlight (photosensitivity) and in some cases even a small amount of sun exposure can cause sunburn or blistering of the skin. Other signs and symptoms often include hearing loss, vision loss, severe tooth decay, bone abnormalities, hands and feet that are constantly cold, and changes in the brain that can be observed on brain scans. Individuals with CS have a serious reaction to an antibiotic medication called *metronidazole*. If affected individuals take this medication it can result in life-threatening liver failure.

CS is sometimes divided into types I, II, and III, based on the severity and age of onset of symptoms. However, the differences between the types are not always clear-cut, and some researchers believe the signs and symptoms reflect a spectrum instead of distinct types. Cockayne syndrome type II is also known as **cerebro-oculo-facio-skeletal (COFS) syndrome**, and while some researchers consider it to be a separate but similar condition, others classify it as part of the Cockayne syndrome disease spectrum.

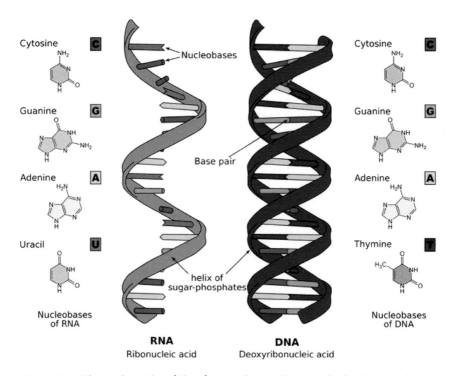

Fig. 7.6. The right side of the figure shows the **two helical strands**. As shown in an earlier figure in the book, DNA contains nucleotides that consist of the bases A, T, C and G that are arranged such that **A** on one DNA strand is always paired with **T** on the other strand and **C** on one strand is always paired with **G** on the other strand. The left side of the figure shows a RNA molecule that is usually **single stranded.**

Base Excision Repair (BER)

A second and distinct excision repair mode called **Base Excision Repair (BER)** is a process during which damaged or inappropriate bases in DNA are removed as the **free base**. Early studies on BER were first initiated to understand details of the removal of the (incorrect) base uracil (U) in DNA. As already noted, **uracil** is normally present in RNA. However, this base can occasionally arise in DNA during DNA replication, and uracil can also appear in DNA following alterations of the chemistry of the bases thymine (T), and cytosine (C) in DNA.

Tomas Lindahl, who in 2015 was awarded the Nobel Prize for his contributions to DNA repair, discovered that when **uracil** is present in DNA, it is removed by an enzyme called **uracil-DNA glycosylase**. **DNA glycosylases** are enzymes that catalyze cleavage of the base-sugar bond of an altered nucleotide residue in DNA, **releasing the damaged base component in free form,** a type of excision repair aptly called **Base Excision Repair** (BER). BER leaves the sugar-phosphate backbone of DNA intact and amenable to further biochemical events required to restore the normal chemistry of the DNA. These events are not described her. Predictably, Uracil-DNA glycosylase removes uracil from DNA, but not from RNA.

BER is not restricted to the repair of uracil in DNA. This story begins when I was inducted into the US Army in the late 1960s. While many of my MD colleagues spent at least a year of their lives in Vietnam, I was fortunate to receive a two-year assignment to the Walter Reed Army Institute of Research (WRAIR) in Washington, DC, where I was presented with a well-equipped research laboratory and a drafted graduate student who served as my research technician. Other than learning how to handle and shoot with an M16 rifle (a weapon used by the US Army in Vietnam) and spending nights downing whiskey at 25 cents a shot in the Walter Reed Officer's Club, sometimes accompanied by a comely drafted female nurse, I discovered an enzyme activity encoded by a bacteriophage (a virus that infects and multiplies in bacteria) called phage T4 that attacked UV-irradiated DNA.

During the course of my two-year stint in the army and subsequently at Stanford University where I was appointed an Assistant Professor in the Department of Pathology, I recognized that the T4 enzyme was a DNA glycosylase that specifically attacks UV-irradiated DNA. I also identified a second enzyme that cuts DNA strands containing sites of base loss, thereby generating apurinic or apyrimidinic (AP) sites.

I and members of my laboratory, notably a graduate student named Eric Radany, eventually learned that the "UV endonuclease" isolated from bacteriophage T4-infected *E. coli* consists of two enzymatic activities that act on UV-irradiated DNA. It was ultimately realized that the phage T4 "UV endonuclease" repairs pyrimidine dimers in strains of *E. coli* defective in excising pyrimidine dimers from DNA. Specifically, a **pyrimidine dimer-DNA glycosylase** excises the dimerized pyrimidines from an affected DNA strand, generating sites of base loss called apyrimidinic sites, leaving a gap (missing two pyrimidines) in the affected DNA strand. Subsequent repair of the single-strand gap by DNA polymerase and DNA ligase activities complete the repair, as they do during conventional nucleotide excision repair (NER) described in the previous chapter (Fig 7.7).

As mentioned earlier uracil belongs to the chemical family of **purines**. When it is removed from DNA by uracil-DNA

THYMINE GLYCOSYLASE/ENDONUCLEASE

**A SEGMENT OF DNA WITH TWO
ADJACENT THYMINE BASES (T T)**

ooooo**T T**ooooo
oooooAAoooooo

**FOLLOWING EXPOSURE TO UV LIGHT THE
T T BASES JOIN TO FORM A THYMINE DIMER (TT)**

ooooo**TT**oooooo
oooooAAoooooo

**THE THYMINE DIMER IS EXCISED TOGETHER
WITH THE RESULTING APYRIMIDINIC SITE
AND THE AFFECTED DNA STRAND IS MISSING
TWO NUCLEOTIDES**
TT

ooooo oooooo
oooooAAoooooo

**THE TWO NUCLEOTIDE GAP IS REPAIRED
BY DNA SYNTHESIS AND JOINED TO THE
EXTANT DNA BY DNA POLYMERASE AND
DNA LIGASE**

ooooo**T T**oooooo
oooooAAoooooo

Fig. 7.7. Excision of thymine dimers by phage T4.

Fig. 7.8. Tomas Lindahl.

glycosylase a site of base loss called an **apurinic (AP) site** is generated in the DNA, thereby essentially exchanging one type of DNA damage for another. Similarly, when the pyrimidines T and C are lost from DNA as free bases, the sites of base loss are called **apyrimidinic sites** (also abbreviated to AP sites), sometimes causing confusion in the literature! AP sites, whether generated by loss of purines or pyrimidines as free bases, are subsequently repaired by a process during which the chemical bond linking the site of base loss to the flanking normal DNA is repaired by a set of reactions that are outside the province of this book.

The occurrence of BER of uracil offers a cogent explanation why DNA does not normally contain **uracil** instead of **thymine**. If it did, the enzyme uracil-DNA glycosylase would presumably not distinguish between uracil that arises by DNA damage from that normally present in DNA, and would essentially destroy the genome.

A final word about base excision repair (BER) is that the process is not restricted to the repair of uracil in DNA. It also participates in the repair of certain types of chemical damage to DNA.

It is now firmly established that the generality of excision repair embraces a wide variety of chemical damage to DNA, much of which involves chemicals that are proven mutagens (agents that cause mutations) and carcinogens (chemicals that cause cancer).

Once the phenomenon of excision repair was established, the challenge was for biochemists to identify and characterize the specific enzymes involved. This proved to be a daunting task. Part of the limitation was the biochemical complexity of excision repair, first hinted at Paul Howard-Flanders's identification of the involvement of three genes. More devilish from the biochemist's point of view was the fact that the protein products of these genes (called *uvrA, uvrB and uvrC*) are constitutively expressed in tiny amounts in *E. coli* and it required the enormously amplifying power of recombinant DNA technology to solve this limitation. Though this solution was late in coming, its ultimate arrival ushered in a new golden era in the history of DNA repair that I'll leave for a future historian to recount.

Chapter 8
Recombinational DNA Repair

When DNA is exchanged between two different chromosomes or between different regions in the same chromosome, the process is referred to as **DNA** or **genetic recombination**. Genetic recombination occurs in both **eukaryotes** (animals and plants) and **prokaryotes** (bacteria). In most situations, in order for an exchange to occur between two chromosomes, the nucleotide (base) sequences containing the swapped regions have to be identical (**homologous**) or similar to a significant degree. Hence, this form of DNA swapping (recombination) is also referred to as **homologous recombination**.

Homologous (genetic) recombination was first described in the bacterium *Escherichia coli* (*E. coli*) in the mid-1940s. For many years, the process in eukaryotes was thought to be the exclusive result of one during which DNA molecules are swapped between those in sperm and in female ova (eggs). However, the process is not restricted to germ cells. Homologous recombination has been observed in all organisms examined from bacteria to man.

One of the notable examples of DNA recombination in cells transpires during a process called **meiosis** (distinct from **mitosis** that you may have learned refers to the division of a cell to generate two daughter cells). During meiosis, **homologous**

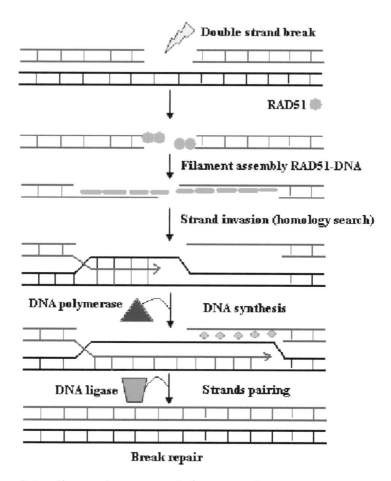

Fig. 8.1. Shows the essential features of recombinational repair in mammalian cells. A break occurs in both strands of a DNA molecule (**red DNA molecule**) that is aligned with a homologous DNA molecule (**black DNA molecule**). A protein called RAD51 binds to the ends of the double-strand break and a **DNA-protein filament** is generated. The normal double-stranded DNA (**black DNA molecule**) is invaded by a strand of the broken DNA (**red DNA molecule**), which seeks DNA sequence homology with a strand of the normal double-stranded DNA (**black DNA molecule**). DNA synthesis by DNA polymerase (**blue triangle**) repairs gaps in both strands of the broken DNA (**red DNA molecule**) and DNA ligase (**orange quadrangle**) ligates the newly synthesized DNA to the extant (preexisting) DNA, thereby completing repair of the double-strand break in the **red DNA molecule**.

chromosomes (chromosome pairs, one from each parent that are the same in length and gene position) swap segments of DNA, usually in germ cells [sperm in males and eggs (ova)].

DNA recombination between homologous parental chromosomes during meiosis generates chromosomes in embryos that are similar but not identical to those of its parents, which is why one's children may resemble their parents in various features but are not identical to them.

Relevant to the topic of DNA repair, homologous recombination is used by cells to repair harmful breaks in one or both DNA strands (the latter being referred to as double-strand breaks (DSB) that occur spontaneously or following some sort of DNA damage.

During recombinational repair, the correct nucleotide (base) sequence is restored by DNA replication using undamaged homologous DNA as a template. Recombinational DNA repair is critical for the survival of UV radiation-damaged cells that accounts for about 50% of the survival of UV irradiated *Escherichia coli*. It is a complicated process that requires two DNA duplexes and the exchange of a strand of DNA from one DNA duplex to the other.

Chapter 9

Mismatch Repair

With respect to the origin of mispaired nucleotides you will hopefully recall from an earlier chapter that during the synthesis (manufacture) of new DNA, the enzyme called DNA polymerase facilitates the incorporation of new nucleotides into a growing DNA strand. This is generally an accurate process during which **A** is incorporated opposite T, **T** is incorporated opposite **A,** **C** opposite **G** and **G** opposite **C**. However, DNA polymerase sometimes makes mistakes and the replication machinery supports the incorporation of an incorrect nucleotide. For example, the nucleotide (base) **A** may be incorporated opposite **G**, **C** or **A** instead of opposite **T** in the new DNA strand.

These misincorporation events transpire with respect to any of the four nucleotides in the DNA strand that are being copied by DNA polymerase. A mismatched base pair can generate a mutation that one can think of as a typographical error in the DNA sequence of the new strand but which can generate a faulty protein that may result in disease.

DNA mismatch repair (MMR) is a DNA repair system for recognizing and repairing erroneous insertion, deletion or incorporation of incorrect nucleotides during DNA replication or during the repair of other types of DNA damage. Incorrect nucleotides in DNA can also arise during DNA recombination.

During mismatch repair two critical events transpire. **The repair machinery must both recognize the mismatched nucleotide pair and also discriminate between the parental DNA strand that is normal and the newly synthesized DNA strand that contains the incorrect nucleotide.** This discrimination is achieved because during normal DNA synthesis the parental DNA strand is chemically modified by the presence of methyl groups, while the newly synthesized strand is initially not methylated. Accordingly, the mismatch machinery "knows" that it must remove the nucleotide from the **unmethylated** DNA strand. The MMR machinery involves multiple proteins and is a biochemically complex process.

MMR is a highly conserved process from bacteria to humans. In fact the first evidence for mismatch repair was obtained from studies in bacteria. Like most, if not all, DNA repair processes, this repair mechanism is highly conserved from bacteria to humans. Research carried out primarily in the bacterium *E. coli*, identified the requirement for multiple genes during mismatch repair, that when inactivated result in bacterial strains that are **hypermutable**, i.e., the strains acquire multiple mutations during DNA replication. The mismatch repair gene products, called Mut proteins, are the active components of the mismatch repair system.

Predictably, loss of normal MMR results in greatly increased rates of spontaneous mutation in organisms ranging from bacteria to humans. In humans, defective MMR often results in hereditary **colorectal cancer**, and is also associated with a significant fraction of sporadic cancers. Given its prominence in mutation avoidance and its ability to target a range of DNA lesions, MMR has been under investigation in studies of ageing

Fig. 9.1. Paul Modrich.

mechanisms. In 2015, Paul Modrich, a leading investigator in the mismatch repair field, together with Tomas Lindahl and Aziz Sancar, were awarded the Nobel Prize in Physiology or Medicine for their contributions on DNA repair.

Chapter 10

The Adaptive and SOS Responses

The Adaptive Response to Alkylation Damage

In addition to the DNA repair pathway called **Base Excision Repair** discussed earlier that can excise multiple alkylated lesions in DNA, two other DNA-repair pathways have been shown to combat the deleterious effects of alkylating agents. In one of these, enzymes called **alkylated guanine transferases (AGTs)** have been shown to reverse certain alkylated guanines, and in a second repair pathway, members of a family of enzymes called **dioxygenases** (enzymes that catalyze the transfer of electrons from one molecule (the **reductant**), to another molecule (the **oxidant**), have been shown to reverse certain alkylated bases that interfere with normal base pairing in DNA.

Alkylating agents comprise a diverse group of chemical compounds. Sources include industrial chemicals, environmental contaminants, naturally occurring compounds, chemotherapeutic drugs and experimental carcinogens.

The penchant of alkylating agents for causing cell death as a consequence of DNA damage led to their use as chemotherapeutic drugs in the treatment of cancer. Scientists involved in DNA repair have pointed out that the double-edged properties of alkylating agents have imparted important significance to studies on cellular pathways that determine the biological outcome of exposure to them.

A feature of the repair of alkylation damage in DNA referred to as the **adaptive response**, was first delineated by Leona Samson and John Cairns in an article entitled, **A New Pathway for DNA Repair in *Escherichia* coli** published in 1997. In a series of subsequent experiments Cairns, Samson and several postdoctoral fellows in the Cairns laboratory elaborated the essential features of this response. They demonstrated that adaptation was effected by exposure of cells to low levels of alkylating agents, and that the response manifested as an increased resistance (**adaptation**) to cell killing and mutagenesis by **subsequent exposure to higher concentrations of the alkylating agents**.

In their initial experiments Samson and Cairns demonstrated that the "adaptive" phenomenon required protein synthesis, providing the first hint that it was an inducible response.

Fig. 10.1. Leona Samson.

Fig. 10.2. John Cairns.

The details of the complex regulation of this inducible response were later revealed in a series of elegant experiments in the laboratories of Tomas Lindahl in London and Mutsuo Sekiguchi in Japan during the 1980s.

It was subsequently established that the genes that were induced in adapted cells encoded at least two types of DNA repair enzymes. Work in several laboratories showed that the adaptation to killing of cells results from enhanced repair of alkylation damage by a new DNA glycosylase that specifically recognizes alkylated bases and is encoded by a gene called ***alkA***.

The enzyme that provides resistance to the mutagenic effects of alkylation of DNA is encoded by a gene called ***ada*** (for **adaptive**) that turned out to be novel in the DNA repair world. Notably, ada protein is endowed with the ability to specifically recognize O_6-**methylguanine** and O_4-**methylthymine** in DNA. Ada protein removes the methyl groups from these two (and only these two) alkylated bases in DNA, transferring them to a specific amino acid residue in the protein. Methylation of Ada protein activates the enzyme, which consequently can only act once! Hence, ada protein is an example of a class of enzymes called *suicide proteins*. Consequently protection (adaptation) from the mutagenic potential of certain alkylating agents is only as effective as the available supply of Ada protein in a cell.

The adaptive response to alkylation damage is one of the most intriguing examples of DNA repair, at least in the bacterium *E. coli*. **O_6-methylguanine-DNA methyltransferase** is not only a DNA repair protein, but additionally when methylated at a different cysteine residue it assumes the role of a positive regulator of the gene that encodes the enzyme, as well as other genes involved in the adaptive response. In so doing the regulatory response ensures the rapid expression of increased amounts of multiple gene products that collectively effect the repair of different types of alkylation damage to DNA.

Although mechanistically distinct from the splitting of pyrimidine dimers by photoreactivating enzyme, the repair of O_6-methylguanine and the monomerization of pyrimidine dimers in DNA both represent examples of DNA repair by the **direct reversal of base damage**, thereby circumventing the need to excise and replace damaged bases and avoiding the potential for introducing errors in the DNA sequence during the repair events.

The SOS Response to DNA Damage

The **SOS response** is a global response to DNA damage named by Miroslav Radman, a brilliant Croation molecular biologist and a long-standing colleague and friend.

The SOS system involves a key protein called RecA in bacteria and Rad 51 in eukaryotes. The RecA protein is stimulated

Fig. 10.3. Miroslav Radman.

by single-stranded DNA and is involved in the inactivation of a group of genes that repress SOS response genes, thereby inducing the response. The SOS response is an error-prone system that results in mutagenesis. Strictly speaking it is erroneous to refer to the SOS response as a DNA repair mechanism or pathway. The response ***does not repair*** *DNA damage* in the strict sense of that phrase. It rescues cells carrying DNA damage from death at the expense of a mutational load. Fundamentally, the SOS response is a mechanism for **cell survival in the face of DNA damage**, not a mechanism for the **repair** of DNA damage. "Where does the label SOS come from?" I once asked Radman. As recounted in my earlier book *Correcting the Blueprint of Life*, his response was as follows:

> "I come from an island in the Adriatic ocean and my father is a fisherman on the island of Cvar who makes his living on the sea. SOS is the well-known international distress signal to save endangered life on the sea. I viewed

the error-prone DNA replication that is the essence of the SOS response as an emergency response of cells to DNA damage whereby the viability of cells can be restored at the cost of accumulating mutations."

The SOS response is complex and beyond the scope of this book. Briefly stated, the response is one in which following DNA damage the cell cycle (a cycle of stages that cells pass through to allow them to divide and produce new cells) is arrested and mutagenesis is induced, resulting in a huge mutational load.

11 DNA Repair in the Context of Chromatin

All transactions that involve DNA in eukaryotes (organisms whose cells contain nuclei), be they DNA replication, transcription, recombination or repair **do not transpire on "naked" DNA**. Eukaryotic genomes are organized and compacted into a complex of proteins called **chromatin**, established by an essential building block called the **nucleosome**, which is generated from an octamer of four core molecules called **histones** (H2A, H2B, H3, H4) around which the DNA is wrapped.

Additionally, the mammalian genome is organized in specific chromatin structures that are characterized by signatures of histone and DNA modifications, specific histone variants and chromatin-associated factors. Consequently, DNA damage must respond to and overcome this major constraint limiting the access of DNA repair factors that are crucial for efficient recognition and removal of DNA lesions.

Regrettably, the importance of chromatin rearrangements and chromatin dynamics during the process of DNA repair is a complex and poorly understood topic. However, the significance of the presence of chromatin in cells undergoing nucleotide excision repair is underscored by experimental observations indicating that chromatin regions with active nucleotide excision repair are more prone to digestion by enzymes that attack DNA (called nucleases), indicating that these regions are rearranged to a more relaxed state and are therefore more sensitive to nuclease treatment compared to compacted chromatin.

Another breakthrough regarding the question of how cells recognize and counteract DNA lesions in the context of chromatin and how the chromatin landscape is altered to facilitate efficient DNA repair, emerged when human fibroblasts were radiolabeled directly after UV irradiation and transferred to a nonradioactive medium for recovery for different time points. It was observed that the sensitivity of the incorporated nucleotides towards nuclease digestion decreased progressively, depending on the time of recovery after UV irradiation and that after a certain time of recovery, the original level of nuclease resistance was restored in the repaired DNA regions.

These findings have built the first principles of the so-called "access-repair-restore" (ARR) model that highlights the conditions for the dynamic chromatin environment during DNA repair. This model summarizes the transient de-condensation of chromatin structures to facilitate recognition and repair of DNA lesions. After DNA lesions are removed, chromatin is reorganized and compacted to its original state.

Chromatin relaxation, as well as its re-condensation following DNA repair is apparently a highly regulated, energy-consuming process that requires the action of **chromatin remodeling complexes**. One family of remodelers primarily conducts functions related to chromatin accessibility, which includes unwrapping of DNA coiled around nucleosomes, nucleosome re-positioning by the sliding of nucleosomes along the DNA, and partial or complete nucleosome eviction.

While the many questions concerning how DNA repair complexes gain access to sites of damage in DNA in its native cellular state, i.e., in the presence of chromatin, remain to be answered, such questions are unrelated to the mechanism by which various established DNA mechanisms transpire, which is of course at the very mechanistic core of **DNA Repair**.

12 Hereditary Diseases Associated with Defective Responses to DNA Damage

As you now hopefully understand, the link between DNA damage, mutagenesis, and malignant transformation is well established. A logical extension is that a congenital defect in a fundamental DNA repair pathway, such as nucleotide excision repair (NER), would be anticipated to be associated with a pronounced cancer predisposition syndrome. Indeed this is well known to be the case when considering the disease called xeroderma pigmentosum (XP).

Xeroderma Pigmentosum (XP)

The pioneering experiments that established excision repair of DNA by Dick Setlow, Paul Howard-Flanders and Phil Hanawalt **were executed using bacteria**, primarily *E. coli*. The history of the discovery that mammalian cells also execute DNA repair and that mutants defective in this process exist, is interesting.

In the mid-1960s eukaryotic molecular biology was in its infancy and the use of direct techniques for demonstrating excision repair in mammalian cells such as measuring the presence of thymine dimers in the nuclei of cells were flawed with technical problems not encountered with *E. coli*. In bacteria undergoing excision repair, excised DNA fragments containing pyrimidine dimers are small enough that they are soluble in acid, whereas the remaining DNA is acid insoluble. Hence one can examine excision repair by comparing the dimer content in the acid soluble and acid-insoluble fractions of cells.

At a meeting on DNA Repair in Chicago in 1965, Dick Setlow stated: Excision — has been looked for and not found in mammalian cells in tissue culture. The DNA of such cells is organized differently from that of bacterial cells, and it is possible that the excised pieces are large enough to be acid insoluble. More sophisticated DNA-fractionation must be utilized before we reach any firm conclusions for mammalian cells.

Shortly thereafter, Ronald Rasmussen and Robert Painter (Fig. 12.1) at the NASA Ames Research Laboratories near Palo Alto, California carried out experiments using a technique called autoradiography to investigate the effects of ionizing radiation (X-rays) on DNA synthesis in mammalian (including human) cells. Having caught wind of the recent discovery of excision repair in bacteria, Rasmussen and Painter were prompted to examine DNA synthesis in mammalian cells exposed to UV light. Years later, Painter wrote: "What we found was a shock. Not only were all S-phase cells (cells synthesizing DNA in anticipation of division) participating, but all the cells in the culture were synthesizing DNA. This phenomenon was referred to as **unscheduled DNA synthesis** because it was transpiring outside the period that mammalian cells normally synthesize DNA.

Rasmussen's and Painter's observations strongly hinted that repair synthesis and hence presumably excision repair, was an evolutionary conserved mechanism. However, other explanations for these observations were tenable, including the possibility of aberrant reinitiation of normal DNA synthesis as a result of exposure of mammalian cells to UV radiation.

Fig. 12.1.   Robert Painter.

A historic breakthrough of this dilemma transpired when James Cleaver joined the scientific staff of the Laboratory of Radiobiology at the University of California in San Francisco (UCSF). Cleaver later recounted to me:

> "I was working with Painter shortly after he had discovered unscheduled DNA synthesis. We were trying to adapt Hanawalt's method for measuring repair synthesis in mammalian cells. But the driving question was how to obtain UV-sensitive mammalian cell mutants to prove that the repair synthesis was really biologically related to excision repair. We were going to attempt to make such mutants ourselves."
>
> In April of 1967, I saw an article in the *San Francisco Chronicle* by the science writer David Perlman. It was brief report of a clinical meeting — the 48th Annual Meeting of the *American College of Physicians* in San Francisco, and it highlighted a talk on the genetics of human cancer by Henry Lynch (mentioned again later in

this chapter). The article described xeroderma pigmentosum (XP) as a genetic disease with a predisposition to skin cancer and sensitivity to sunlight. I thought to myself: "My word, here are God-given UV-sensitive mutants." So we worked through the dermatology department at UCSF and obtained skin biopsies and cultures of XP cells. We got three cultures from three different patients with XP. Painter and I used the methods we had developed at that time for unscheduled DNA synthesis and Hanawalt's technique for repair replication. The results came up on each of the three right away!"

I became aware of Cleaver's fascinating observations that XP cells are defective in DNA repair after they were published in the journal **NATURE** in 1968. I must frankly admit that my immediate pleasure at this extraordinary link between defective DNA repair and human disease was tainted by an inescapable sense of jealousy that this was the "perfect" discovery for a pathologist (like myself) interested in DNA repair, and that it had been discovered

Fig. 12.2.   James Cleaver.

by a physicist instead! In the fifty years since Cleaver's seminal description of defective excision repair in XP patients progress in our understanding of the molecular pathology of this disease has been nothing short of spectacular and will provide fertile material for future historians interested in this topic.

Bob Haynes chuckled when we talked about Cleaver's discovery of the excision repair defect in xeroderma pigmentosum. I vividly remember saying to several people, "Thank God, now we've got a disease. We'll be able to get money from the NIH (a major research funding agency in the US) because DNA repair is relevant to human disease!" In the same proselytizing vein Joshua Lederberg, then chairman of the *Department of Genetics* at Stanford University wrote an editorial for the *Washington Post* in June 1968, soon after Cleaver published his observations:

> "Excitement about the new biology of DNA has tended to provoke either of two reactions: that little men would soon be synthesized and come swirling out of the laboratories, or that the whole study of molecular biology was mainly of academic importance and we would see little practical impact of it within our lifetime."

XP is a genetic disorder with a decreased ability to repair DNA damage caused by ultraviolet (UV) light by nucleotide excision repair (NER). Symptoms may include severe sunburn after only a few minutes in the sun, freckling in sun-exposed areas, dry skin, and changes in skin pigmentation. Nervous system problems, such as hearing loss, poor coordination, loss of intellectual function, and seizures, may also occur, but their mechanistic relationship to defective NER is not obvious. Complications of XP include a high risk of skin cancer (with about half having skin cancer by age 10 without preventive efforts, and cataracts. The risk of other cancers is also evident in this hereditary disease).

In most XP subtypes the devastatingly overt >1000-fold elevated risk of developing malignant tumors on sun-exposed areas of the skin is directly attributable to a failure to remove highly mutagenic solar ultraviolet (UV) radiation-induced DNA photoproducts from the genome. In this sense XP represents a paradigm of a DNA repair disorder with a clear pathological link between genotype and phenotype.

The disease affects about 1 in 20,000 in Japan, 1 in 250,000 people in the United States, and 1 in 430,000 in Europe. It occurs equally commonly in males and females.

There is no cure for XP. Treatment involves complete avoidance of exposure to sunlight, including the use of protective clothing, sunscreen, and dark sunglasses when out in the sun. Retinoid creams may help decrease the risk of skin cancer and vitamin-D supplementation is generally required. If skin cancer occurs it is treated in the usual way. Curiously, life expectancy of those with XP is about 30 years less than normal, despite the fact that as far as is known the disease primarily affects the skin. Clearly there is still much to be learned about the varied clinical manifestation of this disease.

Ataxia Telangiectasia (AT)

Ataxia telangiectasia (AT) (also referred to as Louis–Bar syndrome) is a rare human neurovascular disease (a disease affecting the nervous system and blood vessels), and like XP, displays an autosomal recessive pattern of inheritance. Affected individuals, although clinically normal at birth, develop cerebellar ataxia (loss of muscular coordination) and oculocutaneous

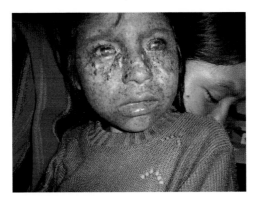

Fig. 12.3. A young girl with xeroderma pigmentosum (XP). Note the obvious damage to the skin of the face, presumably a consequence of exposure to sunlight.

telangiectasia (chronic dilation of the small blood vessels in the eyes and skin) in early childhood. Clinical symptoms include impaired coordination of the torso and/or arms and legs, frequent stumbling, an unsteady gait, uncontrolled or repetitive eye movements, difficulty eating and performing other fine motor tasks, slurred speech, vocal changes and headaches. The disorder is caused by mutations of a gene known as ATM (for "AT Mutated") that has been mapped to chromosome 11.

The course of the disease follows a variable progression commonly leading to total neurological incapacitation before puberty. Accessory complications include bronchiectasis (dilatation of the bronchial tubes), recurrent sino-pulmonary infections (infections in the nose and lungs), impaired cellular immunity and widespread chromosomal instability. AT patients, on receiving conventional radiotherapy for tumor treatment, tend to develop unusually severe complications often culminating in premature death.

Pronounced radiosensitivity (sensitivity to agents such as X-rays) is also observed at the cellular level in laboratory studies, notably the number of radiation-induced chromosomal aberrations is enhanced in white blood cells from AT individuals. Moreover, fibroblasts cultured from affected individuals exhibit a reduced ability to form colonies following exposure to X-rays and radiomimetic chemicals (chemicals that mimic the effects of exposure to X-rays.)

Since the principal damage induced by both types of agents occurs in DNA and seems to be acted on by the same enzymatic repair mechanisms, it would seem probable that the molecular basis for the clinical radiosensitivity of AT patients stems from a deficient DNA repair mechanism. Experiments have measured DNA repair properties of AT fibroblasts after

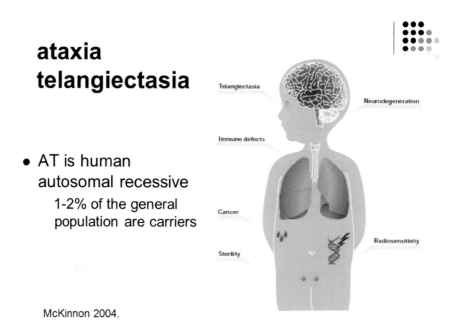

McKinnon 2004.

Fig. 12.4. The cartoon illustrates some of the cardinal clinical features of ataxia telangiectasia (AT).

exposure to X-radiation and have provided direct biochemical evidence that cells from AT donors are impaired in DNA repair; in particular, these cell lines possess an enzymatic defect in an excision-type repair process operating on nitrogenous base residues exposed to X-rays. Curiously, the nature of this postulated excision repair process is unknown and provides an obvious challenge to those interested in the enzymology of DNA repair.

Hereditary Non-Polyposis Colon Cancer (HNPCC) also called Lynch Syndrome

As indicated in the chapter on mismatch repair (MMR), mutations in MMR genes can cause hereditary non-polyposis colorectal cancer (colon cancer without polyps in the colon). Given its prominence in mutation avoidance and its ability to target a range of DNA lesions, MMR has also been under investigation in studies of ageing mechanisms.

Hereditary nonpolyposis colorectal cancer (HNPCC) dates to the description of family studies by Aldrin Scott Warthrin which he began studying in 1895. Warthrin was an American pathologist whose research laid the foundation for understanding the heritability of certain cancers. He has been described as "the father of cancer genetics." The disease is also called Lynch syndrome after Henry T. Lynch, an American physician noted for his discovery of familial susceptibility to certain kinds of cancer and his research into genetic links to cancer. He too is often referred in the literature as "the father of cancer genetics," although Lynch himself has stated that title should belong to Walthrin!

Warthin's observations were not fully appreciated until the mid-1960s when two families with an inheritance pattern of nonpolyposis colorectal cancer and endometrial cancer were described. This condition was first termed the "cancer family syndrome" and was later renamed HNPCC. Some have proposed that HNPCC consists of at least two syndromes: Lynch syndrome I, with hereditary predisposition for colorectal cancer having an early (~44 years) age of onset, a proclivity (70%) for the proximal colon, and an excess of synchronous and asynchronous colonic cancers; and Lynch syndrome II, featuring a similar colonic phenotype (the changes reproducibly observed in an individual with a particular genetic disease) accompanied by a high risk for cancer of the uterus.

Tumors of the ureter and renal pelvis and carcinomas of the stomach, small bowel, ovary, and pancreas also afflict some families. Current estimates indicate that HNPCC may account for as much as 6% of the total colorectal cancer burden.

In addition to xeroderma pigmentosum, ataxia telangiectasia and colorectal cancer associated with defective mismatch repair, there is experimental evidence suggesting that two genetic diseases called **Bloom's Syndrome (BLM)** and **Fanconi's Anemia (FA)** are both causally related to defective responses to DNA damage.

Bloom's Syndrome

Bloom's syndrome (seldom called BS for obvious reasons) is a rare genetic disorder in which the cell's ability to maintain the integrity of DNA is impaired. DNA instability is displayed at the chromosomal level as striking increases in exchanges between chromosomes, as well as increases in chromosome gaps and breaks. The syndrome is inherited in an autosomal recessive manner, wherein both the maternally- and paternally-derived copies of the Bloom's syndrome gene *(BLM)* are functionally deficient.

The condition was first described and discovered by the New York dermatologist Dr. David Bloom in 1954. Bloom's is also known as "Bloom-Torre-Machacek syndrome".

Bloom's syndrome is an inherited disorder characterized by short stature, a skin rash that develops after exposure to the sun, and a greatly increased risk of cancer. People with Bloom's syndrome are usually smaller than 97 percent of the population in both height and weight from birth, and they rarely exceed 5 feet tall in adulthood.

Affected individuals have skin that is sensitive to sun exposure, and they usually develop a butterfly-shaped patch of reddened skin across the nose and cheeks. A skin rash can also appear on other areas that are typically exposed to the sun, such as the back of the hands and the forearms. Small clusters of enlarged blood vessels (telangiectasia) often appear in the rash; telangiectasia can also occur in the eyes. Other skin features include patches of skin that are lighter or darker than the surrounding areas (hypopigmentation or hyperpigmentation, respectively). These patches appear on areas of the skin that are not exposed to the sun, and their development is not related to the rashes.

Individuals with Bloom's syndrome have a high-pitched voice and distinctive facial features, including a long, narrow face, a small lower jaw and a prominent nose and ears. Other features can include learning disabilities, an increased risk of diabetes, chronic obstructive pulmonary disease (COPD), and mild immune system abnormalities leading to recurrent infections of the upper respiratory tract, ears, and lungs during infancy. Men with Bloom's syndrome usually do not produce

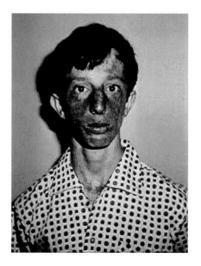

Fig. 12.5.   Bloom's syndrome.

sperm and as a result are unable to father children. Women with the disorder generally have reduced fertility and experience menopause at an earlier age than usual.

Mutations in the BLM gene lead to this syndrome. The BLM gene offers directions for producing a family of proteins called the RecQ helicases. Helicases are enzymes that adhere to DNA and unwind the two spiral strands of the DNA molecule. The unwinding of the strands is essential for copying DNA in preparation for cell division.

If the *BLM* gene is mutated, BLM protein is hindered from performing its function of ensuring genomic stability. This leads to an increase of sister chromatid exchange to 10-fold. This chromosomal instability causes gaps and breaks in DNA, which leads to defects in normal cell activities. BS is inherited in an autosomal recessive pattern, implying that both copies of the gene in each cell have alterations. Parents of a child with an autosomal recessive condition carry one copy of the altered gene but usually do not show symptoms of the condition.

There is no treatment available to make the genome in the cells stable and block mutations. However, children who have poor feeding habits and have been suffering from gastro-esophageal reflux require non-volitional feeding. The facial skin patches have to be protected from sunlight by avoiding exposure and the use of head cover and sunscreens.

Fanconi Anemia

Fanconi anemia (FA) is a rare genetic disease resulting in impaired response to DNA damage. Although very rare, studies of this and other bone marrow failure syndromes have improved our understanding of the mechanisms of normal bone marrow function and development of cancer. Among those affected with FA, the majority develop cancer, most often acute myelogenous leukemia, and 90% develop bone marrow failure

Fig. 12.6. Guido Fanconi.

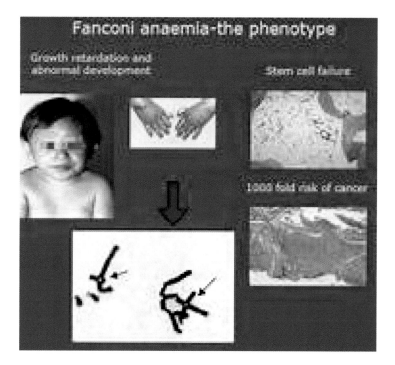

Fig. 12.7.

(the inability to produce blood cells) by age 40. About 60–75% of people have congenital defects, commonly short stature, abnormalities of the skin, arms, head, eyes, kidneys, and ears, and developmental disabilities. Around 75% of people have some form of endocrine problems with varying degrees of severity.

FA is the result of a genetic defect in a cluster of proteins responsible for DNA repair via homologous recombination. Treatment with androgens (male sex hormones such as testosterone) and hematopoietic (blood cell) growth factors can help bone marrow failure temporarily, but the long-term treatment is bone marrow transplant if a donor is available. Because of the genetic defect in DNA repair, cells from people with FA are sensitive to drugs that treat cancer by DNA crosslinking such as mitomycin C. The typical age of death was 30 years in 2000.

FA occurs in about one per 130,000 births, with a higher frequency in Ashkenazi Jews in Israel and Afrikaners in South Africa. The disease is named after Guido Fanconi, a Swiss pediatrician who originally described this disorder,. It should not be confused with **Fanconi syndrome**, a kidney disorder also named after Fanconi.

Chapter 13   Conclusion

I hope that you the reader enjoyed and learned something from reading **DNA Repair** and that you now have a reasonable idea of what the repair of damage to our genes is about and how impossible life would be in the absence of these biological mechanisms. The comprehension that multiple diverse DNA repair mechanisms that emerged during the eons of biological evolution clearly indicates the importance of healthy genes in order to protect us and other living species from genetically inherited diseases at different times and places during evolution.

For example, the existence of the adaptive response to DNA damage presumably reflects the prevalence of DNA damage by alkylating agents as a primary source of DNA damage at a particular time and place or places, an observation not easily understood except to suggest that multiple varying types of DNA damage at different times and places on planet earth drove the evolutionary **selection** of different modes of DNA repair. Note the significant distinction between **selection** and **emergence**. One must assume that multiple DNA repair modes evolved before their utility was evident and were selected once they were beneficial for survival.

The diversity of the known repertoire of DNA repair mechanisms is also instructive and presumably reflects the diversity of DNA damage. I hope, too, that you now recognize that DNA repair is an indispensible biological function without which the frequency of mutations in our genes would be horrendous, leading to all manner of diseases and disabilities, many of which, including cancer, might have made life on planet Earth unsustainable.

If you have not gained sufficient knowledge of genes and DNA — what they are, where they reside in our cells, and how they function, I invite you to examine a companion book published earlier in 2018 by World Scientific Publishing, entitled, **Learning About Your Genes — A Primer for Non-Biologists**. Readers with a more comprehensive understanding of molecular biology in general and DNA repair in specific, who are interested in gaining more advanced information, may want to examine **Correcting the Blueprint of Life**. You should also feel free to contact me at:

errol.Friedberg@utsouthwestern.edu
or at
errolfriedberg@gmail.com

Glossary

A

adduct in biology, an adduct is a complex that forms when a chemical binds to a biological molecule, such as DNA or protein

agar plates glass or plastic plates containing agar, on which bacteria grow

Albert Kelner a scientist who discovered a DNA repair mode called photoreactivation

Alexander Fleming a British scientist who discovered the antibiotic, penicillin

Alexander Hollaender Alexander Hollaender was one of the leading researchers in radiation biology and in genetic mutations. In 1983, he was given the Enrico Fermi Award by the United States Department of Energy for his contributions in founding the science of radiation biology, and for his leadership in promoting scientific exchanges between scientists and scientists from developing countries

alkylated bases	alkylated bases in DNA or RNA that have been exposed to chemicals that add alkyl groups
alkylguanine transferase	O^6-alkylguanine DNA transferase is a DNA repair protein that protects cells from killing and mutagenesis by alkylating agents
antibiotics	antimicrobial substances in active bacteria
anti-parallel	the orientation of parallel molecules in opposite directions, such as in double-stranded DNA
apoptosis	the death of cells which occurs as a normal and controlled part of an organism's growth or development
ataxia telangiectasia	ataxia telangiectasia (or AT), also referred to as Louis–Bar syndrome, is a rare neurodegenerative autosomal recessive disease that causes severe disability
autoradiography	a photographic technique used to localize a radioactive substance within a solid specimen
autosome	any chromosome that is not a sex chromosome
Arthur Kornberg	Arthur Kornberg was an American scientist who won the Nobel Prize in Physiology or Medicine for his discovery of the biological synthesis of DNA
Aziz Sancar	a Turkish-American biochemist and molecular biologist in DNA repair, cell cycle checkpoints, and circadian clock. In 2015, he was awarded the Nobel Prize in Chemistry

B

bacterium	a microscopic life form
base excision repair (BER)	a mode of DNA repair during which damaged bases are removed as free bases
base pair	a pair of two facing bases in opposite DNA strands
Bloom's syndrome	(sometimes abbreviated as BS in the literature), also known as Bloom-Torre-Machacek syndrome, is a rare autosomal recessive disorder characterized by short stature, predisposition to the development of cancer and genomic instability
buffer	a solution that resists changes in pH when acid or alkali is added to it. Buffers typically involve a weak acid or alkali together with one of its salts

C

Claude Stan Rupert	a scientist who conducted seminal research in the field of light-activated DNA repair
chromatin	the material of which the chromosomes of organisms other than bacteria (i.e., eukaryotes) are composed. It consists of protein, RNA, and DNA
chromatin remodeling	Chromatin remodeling is the dynamic modification of chromatin architecture to allow access of condensed genomic DNA to the regulatory transcription machinery proteins, and thereby control gene expression

chromosome   a rod-shaped structure carrying DNA found in the nucleus of cells

chromosome pair   in humans, each cell normally contains 23 pairs of chromosomes for a total of 46. 22 of these pairs look the same in both males females. The 23rd pair of chromosomes are called X (in females) and Y (in males)

Claud Rupert   Claud Stanley Rupert conducted seminal research in the field of light-activated DNA repair and was one of the founding faculty members of The University of Texas at Dallas

Cold Spring Harbor   founded in 1980, Cold Spring Laboratory Harbor Laboratory has shaped contemporary biomedical research and education with programs in cancer biology, neuroscience, plant biology and quantitative biology

cyclobutane pyrimidine dimers   pyrimidine dimers are molecular lesions formed from thymine or cytosine bases in DNA via photochemical reactions

cytosine   a base found in DNA and RNA. It is paired with guanine in double-stranded DNA

D

deoxyribose   a sugar with five carbon atoms derived from the pentose sugar ribose that is found in nucleotides

deoxyribonucleic acid DNA   a molecule that encodes an organism's genetic blueprint; in other words, DNA contains all of the information required to build and maintain an organism

DNA damage an alteration in the chemical structure of DNA, such as a break in a strand of DNA, a base missing from the backbone of DNA, or a chemically altered base. DNA damage can occur naturally (endogenous DNA damage) or via environmental factors (exogenous DNA damage)

DNA glycosylase DNA glycosylases are a family of enzymes involved in base excision repair of DNA

DNA polymerase DNA polymerase is an enzyme that synthesizes DNA from deoxyribonucleotides, the building blocks of DNA

DNA recombination DNA recombination is the exchange of DNA strands to produce new nucleotide sequence arrangements. Recombination occurs typically, though not exclusively, between regions of similar sequence by breaking and rejoining DNA segments, and is essential for generating genetic diversity and for maintaining genome integrity

DNA repair a collection of processes by which cells identify and correct damage to the DNA molecules that encode its genome

DNA replication DNA replication is the biological process of producing two identical replicas of DNA from one original DNA molecule. This process occurs in all living organisms and is the basis for biological inheritance

DNA sequence a DNA sequence is the order of any given number of adjacent nucleotides in DNA

DNA sequencing

the complete sequence of a DNA strand can be obtained by a technique called DNA sequencing. DNA sequencing is a process of determining the accurate order of nucleotides along chromosomes and genomes. It includes any method or technology that is used to determine the order of the four bases — adenine, guanine, cytosine, and thymine in a strand of DNA

DNA template

a DNA template is a single-stranded molecule of DNA that directs the correct nucleotide sequence of the complementary strand during DNA replication

DNA transcription

Transcription is the process by which the information in a strand of DNA is copied into a new molecule of messenger RNA (mRNA)

Drosophila melanogaster

Drosophila melanogaster has been extensively studied for over a century as a model organism for genetic investigations. It also has many characteristics that make it an ideal organism for the study of animal development and behavior, neurobiology, and human genetic diseases

E

enzyme

a substance produced by a living organism that acts as a catalyst to bring about a specific biochemical reaction

Escherichia coli

Escherichia coli (abbreviated as *E.coli*) are bacteria found in the environment, foods, and intestines of people

eukaryote   an organism consisting of a cell or cells in
                       which the genetic material is DNA in the form
                       of chromosomes contained within a distinct
                       nucleus. Eukaryotes include all living organ-
                       isms other than the eubacteria and archae-
                       bacteria; microorganisms that are similar to
                       bacteria in size and simplicity of structure but
                       radically different in molecular organization

excision repair   excision repair is any type of DNA repair
                       during which damaged bases or nucleotides
                       in DNA are removed

F

Fanconi anemia   a rare disease passed down through families
                       (inherited). It results in decreased produc-
                       tion of all types of blood cells. It is the most
                       commonly inherited form of aplastic anemia

Fibroblast   a fibroblast is a type of biological cell that
                       synthesizes the extracellular matrix and col-
                       lagen, produces the structural framework
                       for animal tissues, and plays a critical role
                       in wound healing. *Fibroblasts* are the most
                       common cells of connective tissues in animals

fluorescent light an electric current in the gas excites mercury
                       vapor, which produces short-wave ultraviolet
                       light that then causes a phosphor coating
                       on the inside of the lamp to glow. A fluo-
                       rescent lamp converts electrical energy into
                       useful light much more efficiently than an
                       incandescent lamp

Francis Crick   Francis Harry Compton (1916–2004) was a biophysicist and neurobiologist most noted as the co-discoverer of the structure of DNA in 1953. With James Watson and Maurice Wilkins, he was jointly awarded the 1962 Prize in Physiology or Medicine for their discoveries concerning the molecular structure of DNA

Franklin (Frank) Stahl   Franklin (Frank) William Stahl is an American molecular biologist and geneticist who conducted the famous Meselson-Stahl experiment showing that DNA is replicated by a semiconservative mechanism-meaning that each strand of DNA serves as a template for synthesis of a new strand

G

gene   (in informal use) a unit of heredity that is transferred from a parent to offspring and is held to determine some characteristics of the offspring. In technical use, a distinct sequence of nucleotides forming part of a chromosome, the order of which determines the order of monomers in a polypeptide or nucleic acid molecule that a cell (or virus) may synthesize. The genetic properties or features of an organism's characteristics, etc.

genetic code   the genetic code is the set of rules used by living cells to translate information encoded within genetic material (DNA or messenger RNA sequences) into proteins

geneticist   an expert in or student of heredity and the variation of inherited characteristics during protein synthesis

genetic material   genetic material is called DNA and/or RNA. DNA is the hereditary material found in the nucleus of eukaryotic cells (animal and plant) and the cytoplasm of prokaryotic cells (bacteria) that determines the composition of an organism

genetic recombination   DNA recombination involves the exchange of genetic material either between multiple chromosomes or between different regions of the same chromosome

genome   the complete set of genes or genetic material present in a cell or organism

genomics   the branch of molecular biology concerned with the structure, function, evolution and mapping of genes

gastro-esophageal reflux   gastroesophageal reflux disease (GERD) is a clinical manifestation of the excessive reflux of acidic gastric contents into the esophagus causing various degrees of symptomatic irritation or injury to the esophageal mucosa

genomics   the branch of molecular biology concerned with the structure, function, evolution, and mapping of genomes

germ cell   a germ cell is any biological cell that gives rise to the gametes (a gamete is a mature haploid male or female germ cell which is able to unite with another of the opposite sex in sexual reproduction to form a diploid cell called a zygote) of an organism that reproduces sexually. In many animals, the germ cells in the primitive streak of an embryo migrate via the gut to the developing gonads

guanine   a compound that occurs in guano and fish scales, and is one of the four constituent bases of nucleic acids. A purine derivative, it is paired with cytosine in double-stranded DNA

Guido Fanconi   Guido Fanconi was a Swiss pediatrician. He was born in a small village in the Canton of Grisons in Switzerland and is regarded as one of the founders of modern pediatrics

H

Henry Lynch   Henry T. Lynch was an American physician noted for his discovery of familial susceptibility to certain kinds of cancer and his research into genetic links to cancer

hereditary material   a gene is a short piece of DNA (deoxyribonucleic acid), the hereditary material in humans and almost all other organisms. There are about 30,000 genes in each cell of the human body

Hermann Muller	an American geneticist best remembered for his demonstration that mutations and hereditary changes can be caused by X-rays striking the genes of living cells. His discovery of artificially induced mutations in genes had far-reaching consequences, and he was awarded the Nobel Prize in 1946
histones	any of a group of basic proteins found in chromatin
homologous recombination	Homologous recombination is a type of genetic recombination in which nucleotide sequences are exchanged between two similar or identical molecules of DNA. It is most widely used by cells to accurately repair harmful breaks that occur on both strands of DNA, known as double-strand breaks (DSB)
Hvar	a Croatian island in the Adriatic Sea, best known as a summer resort

I

ionizing radiation	ionizing radiation affects living things on an atomic level, by ionizing molecules in cells

J

James Cleaver	James E. Cleaver is Emeritus Professor of Dermatology and Pharmaceutical Chemistry in the Department of Dermatology at the University of California San Francisco (UCSF). In 1968, he discovered that the hereditary human skin cancer in people with xeroderma pigmentosum is caused by mutations in genes that code for DNA repair

James Watson	James Watson is a molecular biologist who together with Francis Crick solved the structure of DNA in 1953 and until recently was the retired director of the Cold Spring Harbor Laboratory
John Cairns	John Cairns was a British physician and molecular biologist who made significant contributions to molecular genetics and cancer research
Johns Hopkins University	founded in 1876, it is America's first research university and home to nine world-class academic divisions working together as one university
Julius Petri	Petri's invention — a shallow dish used to grow and identify bacterial strains, revolutionized the world of microbiology and the way we culture microorganisms

L

Leona Samson	Leona Samson is the Uncas and Helen Whitaker Professor and American Cancer Society Research Professor at the MIT. Her research interests focus on methods for measuring DNA repair capacity in human cells
liquid holding recovery	a term first used to describe the survival of UV-irradiated cells in a buffer solution, during which recovery required from 48 to more than 72 hours at 22°C, while photoreactivation was completed within 4–6 hours

M

methylthymine	methylthymine is a minor lesion that is induced by DNA methylating agents

Milislav Demerec	a Croatian-American geneticist, and director of the Department of Genetics at the Carnegie Institution of Washington (now the Cold Spring Harbor Laboratory), from 1941 to 1960
Miroslav Radman	a geneticist and molecular biologist recognized for his groundbreaking work on DNA repair, recombination and mutation and their impact on biological evolution and human health
mismatch repair (MMR)	DNA mismatch repair (MMR) is a system for recognizing and repairing erroneous insertion, deletion, and misincorporation of bases that can arise during DNA replication and recombination, as well as the repair of some forms of DNA damage
molecular biology	the study of biology at the molecular level; the branch of biology that deals with the structure and function of macromolecules (e.g., proteins and nucleic acids) essential to life
mutagen	a physical or chemical agent that changes the genetic material, usually thus increasing the frequency of mutations above the natural background level
mutagenic	capable of inducing mutation or increasing its rate
mutation	the changing of the structure of a gene, resulting in a variant form that may be transmitted to subsequent generations, caused by the alteration of single base units in DNA or the deletion, insertion, or rearrangement of larger sections of genes or chromosomes

Mutsuo Sekiguchi

Mutsuo Sekiguchi and his colleagues independently discovered phage T4 endonuclease V (five), a DNA repair enzyme that functions to ultraviolet- damaged DNA

N

***Nature* magazine**

Nature is a British multidisciplinary scientific journal first published in 1869. It was ranked the world's most cited scientific journal by the Science Edition of the 2010 Journal Citation Reports and is ascribed an impact factor of 40.137, making it one of the world's top academic journals

Nobel Prize

any of six international prizes awarded annually for outstanding work in physics, chemistry, physiology or medicine, literature, economics (since 1969), and the promotion of peace. The Nobel Prizes, first awarded in 1901, were established by the will of the Norwegian Parliament and by a board of deputies appointed by Swedish learned societies

nucleosome

a nucleosome is a basic unit of DNA packaging in eukaryotes, consisting of a segment of DNA wound in sequence around eight histone protein cores. This structure is often compared to thread wrapped around a spool

nucleotide an organic molecule that is the building block of DNA and RNA and is made up of three parts: a phosphate group, a 5-carbon sugar, and a nitrogenous base. The four nitrogenous bases in DNA are adenine, cytosine, guanine, and thymine. RNA contains uracil instead of thymine

nucleotide excision repair nucleotide excision repair (NER) is a particularly important excision mechanism that removes DNA damage induced by ultraviolet (UV) light or by chemicals that can bind to DNA

O

oxidant an oxidant is a reactant that oxidizes or removes electrons from other reactants during a redox reaction. An oxidant may also be called an oxidizer or oxidizing agent

P

Paul Modrich an American biochemist who discovered mismatch repair, a mechanism by which cells detect and correct errors that are introduced into DNA during DNA replication and recombination. Modrich was among the first to show that a common form of inherited colorectal cancer is due to defective mismatch repair. For his contributions, Modrich received the Nobel Prize in 2015

phage   short for *bacteriophage*, a virus that lives within bacteria. A virus for which the natural host is bacterial cells

Philip Hanawalt   an American biologist who discovered the process of repair replication during nucleotide excision repair of damaged DNA and subsequently the process of transcription-coupled excision repair

photoproducts   products of photochemical reactions

photoreactivation   photoreactivation (PR) is the recovery from biological damage caused by UV radiation

photoreactivating enzyme   an enzyme that repairs damage in the presence of visible light caused by exposure to ultraviolet light

polymorphism   polymorphism in biology is a discontinuous genetic variation resulting in the occurrence of several different forms or types of individuals among the members of a single species

prokaryote   a microscopic single-celled organism that has neither a distinct nucleus with a membrane nor other specialized organelles; includes cyanobacteria

protein   any of a class of nitrogenous organic compounds that consist of large molecules composed of one or more long chains of amino acids and are an essential part of all living organisms, especially as structural components of body tissues such as muscle, hair, collagen, etc., and as enzymes and antibodies

purines	in DNA, these bases form hydrogen bonds with their complementary pyrimidines, thymine and cytosine, respectively. In RNA, the complement of adenine is uracil instead of thymine
pyrimidines	in DNA and RNA, these bases form hydrogen bonds with their complementary purines. Thus, in DNA, the purines adenine (A) and guanine (G) pair with the *pyrimidines* thymine (T) and cytosine (C), respectively
pyrimidine-pyrimidone (6–4) products	6–4 photoproducts are photoproducts induced in DNA by exposure to ultraviolet (UV) light
pyrimidine dimers	pyrimidine dimers are molecular lesions formed from thymine or cytosine bases in DNA via photochemical reactions. Ultraviolet light (UV) induces the formation of covalent linkages between consecutive bases along the nucleotide chain in the vicinity of their carbon–carbon double bond

R

radiation	the emission of energy as electromagnetic waves or as moving subatomic particles, especially high-energy particles that cause ionization
radiobiology	radiobiology (also known as radiation biology) is a field of clinical and basic medical sciences that involves the study of the action of ionizing radiation on living things, especially health effects of radiation

recombinational repair	in *Escherichia coli*, two-strand DNA damage, generated mostly during replication on a template DNA containing single-strand damage, is repaired by recombination with a homologous intact duplex, usually the sister chromosome
Renato Dulbecco	an Italian-American, who won the 1975 Nobel Prize in Physiology or Medicine for his work on oncoviruses, viruses that can cause cancer when they infect animal cells. He also co-discovered photoreactivation independent of Albert Kelner
ribonucleic acid (RNA)	ribonucleic acid or RNA is one of the two types of nucleic acids found in life on Earth. The other deoxyribonucleic acid (DNA) has long assumed a higher profile than RNA in the minds of casual observers
ribose	ribose is a carbohydrate; specifically, it is a pentose monosaccharide (simple sugar). Ribose is found in RNA
Richard Setlow	Richard "Dick" Setlow was a biophysicist recognized for his research on DNA damage and repair and a senior scientist emeritus at the U.S. Department of Energy's Brookhaven National Laboratory
RNA polymerase	RNA polymerase (ribonucleic acid polymerase), both abbreviated RNAP or RNA pol, official name DNA-directed RNA polymerase, is a member of a family of enzymes that are essential to life: they are found in all living organisms and viruses

Robert H. Haynes	Robert Hall Haynes was a Canadian geneticist and biophysicist. He was the Distinguished Research Professor in the Department of Biology at York University, Canada
Robert Lehman	is an emeritus professor of biochemistry at the Stanford University School of Medicine who made major contributions in characterizing the process of homologous recombination
Robert Painter	Robert Painter was a professor in the Department of Radiobiology at the University of California, San Francisco, who made seminal contributions to DNA repair
Rockefeller Institute	The Rockefeller University is a center for scientific medical research, primarily in the biological and medical sciences, that provides doctoral and postdoctoral education
Ruth Hill	The isolation of a UV radiation-sensitive mutant of *E. coli* by Ruth Hill in 1958 inaugurated the era of the genetic approach to the study of excision repair of DNA
Ronald Rasmussen	In the mid-1960s, Ronald Rasmussen, together with Robert Painter, demonstrated that UV-irradiated mammalian cells incorporated nucleotides outside the S-phase of the cell cycle that they correctly interpreted as DNA synthesis during DNA repair

S

Salk Institute	the Salk Institute for Biological Studies is an independent, non-profit, scientific research institute located in the La Jolla community in San Diego, California. It was founded in 1960 by Jonas Salk, the developer of the polio vaccine

Salvador Luria   an Italian microbiologist, later a naturalized US citizen. He won the Nobel Prize in1969 with Max Delbrück and Alfred Hershey for their discoveries of the DNA replication mechanism

senescence   the condition or process of deterioration with age — in biology, the loss of a cell's ability to divide and grow

sequestered   isolated and hidden away

silent mutation   a form of point mutation resulting in a codon that codes for the same or a different amino acid, but without any functional change in the protein product

single-stranded DNA   a DNA molecule consisting of only a single helical form. In nature, single-stranded DNA genomes can be found in *parvoviridae* (class II viruses)

spontaneous mutation   a naturally occurring mutation in the absence of a mutagen that would otherwise be a known factor for inducing a particular mutation

Stanford University   a private research university in California. Stanford is known for its academic strength, wealth, proximity to Silicon Valley, and ranking as one of the world's top universities

Staphylococcus   *Staphylococcus aureus* (or Staph) is a type of bacteria commonly found on the skin and hair as well as in the nose and throat of people and animals. These bacteria are present in up to 25 percent of healthy people and are more common among those with skin, eye, nose, or throat infections

T

thymine a substance that is one of the four constituent bases of nucleic acids. A pyrimidine derivative, it is paired with adenine in double-stranded DNA

Tomas Lindahl a Swedish-born British scientist who among other contributions discovered a unique class of enzymes in mammalian cells, namely the methyltransferases that mediate the adaptive response to DNA damage. Lindahl was awarded the Nobel prize in 2015

transcription DNA transcription is a process that involves transcribing genetic information from DNA to RNA. The transcribed DNA message, or RNA transcript, is used to produce proteins. DNA is housed within the nucleus of cells

transcription-coupled DNA repair transcription-coupled repair is a subpathway of NER dedicated to the repair of lesions that by virtue of their location on the transcribed strands of active genes, encumber elongation by RNA polymerases

transforming principle a term used by scientists in the 1930s and 1940s before they had isolated DNA and identified it as the actual transforming substance

tumor suppressor genes a type of gene that makes a protein called a tumor suppressor protein that helps control cell growth. Mutations in tumor suppressor genes may lead to cancer. Also called antioncogenes

U

ultraviolet radiation (light)	electromagnetic radiation with a wavelength from 10 nm to 400 nm; shorter than that of visible light but longer than X-rays
uracil	a pyrimidine base that is one of the fundamental components of RNA, in which it forms base pairs with adenine
US National Academy of Science	a United States non-profit governmental organization. New members are elected by current members based on their distinguished and continuing achievements in original research. Election to the National Academy is one of the highest honors in science

V

virulent	The degree or ability of a pathogenic organism to cause disease

W

Wilhelm Conrad Röntgen	Wilhelm Conrad Röntgen was a Nobel-winning German scientist who is credited with the discovery of X-rays
Wilhelm Johannsen	a Danish botanist, plant physiologist, and geneticist best known for coining the terms gene phenotype and genotype

X

xeroderma pigmentosum	xeroderma pigmentosum (XP) is a genetic disorder in which there is a decreased ability to repair DNA damage caused by ultraviolet (UV) light
X-rays	an electromagnetic wave of high energy and very short wavelength, which is able to pass through many opaque materials

Index